CW00431438

CONCISE
DICTIONARY
OF
SCIENCE

Concise
Dictionary
of
Science

Keith Whittles BSc, PhD, FGS
Alice Goldie BSc

TIGER BOOKS INTERNATIONAL
LONDON

Tiger Books International PLC, London

This edition published in 1993 by
Tiger Books International PLC, London.

ISBN 1-85501-369-X

Printed and bound in Slovenia.

Contents

Contents

A

abbe refractometer a device for measuring the REFRACTIVE INDEX of liquids.

aberration in physics, the result when a lens does not produce a true image. In astronomy, the apparent positional change of a body.

abscissa the horizontal or x-co-ordinate in GEOMETRY that is the distance of the point from the vertical or y-axis. For example, a point with CARTESIAN CO-ORDINATES (2,–6) has an abscissa of 2.

absolute temperature a temperature measured on the KELVIN SCALE with respect to ABSOLUTE ZERO.

absolute zero the temperature at which the particles of matter have no thermodynamic energy, theoretically given the lowest value of –273.15°C (–459.67°F).

ABO blood group *see* **blood grouping**.

AC *see* **alternating current**.

acacia gum *see* **gum arabic**.

acceleration the rate of change of velocity with time. It is usually denoted by the symbol a and has SI units of metres per second squared (ms⁻²). The acceleration of an object can be calculated using DIFFERENTIATION or the equation:

$a = F/M$ where F = force and M = mass.

Independent of their mass, all bodies falling freely under the force of GRAVITY do so with uniform acceleration, particularly if air resistance is negligible.

Acceleration due to gravity is approximately 9.8 ms^{-2} (32.17ft), thus where air resistance is negligible, the velocity of a falling object increases by 9.8 ms^{-1} per second. On the other hand, an object shot straight upwards with a particular velocity will decelerate by 9.8 ms^{-1} every second until it reaches its maximum height after a specific period of time. When its maximum height is reached, the object that had been shot straight upwards (no horizontal motion whatsoever) will start to fall with an acceleration of 9.8 ms^{-2}.

accelerator a substance used to increase the rate of a chemical reaction (*see* CATALYST), or beneficially affect the action of an enzyme.

acceptor an ATOM that accepts ELECTRONS in forming a bond *or* reactant A in an induced reaction using inductor B, where the reaction rate with reactant C is increased by the presence of INDUCTOR B. In electronics, an impurity which causes HOLE conduction in a semiconductor through the movement of electrons to the impurity, creating holes, or positive charge carriers.

accretion in geology, the addition of material to the edge of a continent. Similarly, when applied to a celestial body, e.g. a planet, the process of accumulation of matter onto the body, driven by the influence of GRAVITY.

acetic acid, more correctly ethanoic acid (CH_3COOH), is obtained by the oxidation of ethanol or synthesized from acetylene. It is the acid in vinegar (an aqueous solution of 3 to 6 per cent).

acetone or propanone (dimethyl ketone CH_3COCH_3) is an important solvent used much in organic synthesis, e.g. in making plastics.

acetylcholine one of the important substances in the nervous system, transmitting impulses from one nerve to another (*see* SYNAPSE).

acetylene *see* **ethyne**.

acicular pointed or needle shaped. Applied particularly in the description of minerals (*see also* HABIT).

acid any substance that releases hydrogen ions during a chemical reaction, thus lowering the pH of the solution. An acidic solution has a pH of less than 7 and will react with a BASE to form a salt and water. An acid can be thought of as a proton (H+) donor (BRÖNSTED-LOWRY definition) or as an electron pair acceptor (LEWIS definition).

acid rain rain with a high concentration of pollutants, such as dissolved sulphur and nitrogen oxides, which have harmful effects on plant and animal life. The pollutants are principally byproducts of industrial activities involving the burning of coal or oil (*see* FOSSIL FUELS).

acid rocks igneous rocks containing over 66 PER CENT SILICA by weight. Most of the silica is in the form of silicate minerals but some exists, the excess, as free QUARTZ. Rock types include GRANITES and rhyolites (a fine-grained EXTRUSIVE ROCK).

acquired immune deficiency syndrome *see* **AIDS**.

acquired immunity *see* **adaptive immune system, immune system**.

acre a measure of area, equal to 4840 square yards (4047 square metres and 0.4 hectares).

acrylic resins a class of thermoplastics formed by polymerization of acrylic acid (propenoic acid) derivatives, usually AMIDES or ESTERS. They are colourless and transparent and are resistant to most weak chemicals, ageing and light. They are commonly used for optical products, e.g. lenses. Trade names include Perspex.

actin a protein with globular and fibrous (polymerized) forms, found in muscle with MYOSIN, and now known to be associated with sites of cellular movement.

actinides the collective name for elements in the PERIODIC TABLE with ATOMIC NUMBERS from 89 (actinium) to 103 (Lawrencium) inclusive. All the elements resemble actinium and are radioactive, and many do not occur naturally. The actinide metals resemble the LANTHANIDES, are of low ELECTRONEGATIVITY, and are very reactive. The metals are produced by electrolytic reduction of fused salts or by treating the HALIDES with calcium at high temperatures.

action potential a voltage pulse produced in a nerve by a stimulus. A continuing stimulus produces frequent pulses resulting in muscular responses.

activated carbon carbon with significant facility for adsorbing large quantities of gases. Used in gas masks, medicine and in industrial processes to remove solvent vapours.

activation energy the input of energy required to initiate a chemical reaction. In some reactions, the KINETIC ENERGY of the reactant molecules is sufficient for the reaction to proceed, but in others the activation energy is too high for the reaction to occur (*see* CATALYST).

active margin the margin of a continent which is also the edge of a PLATE. Because of the plate tectonic activity, such margins are sites of EARTHQUAKES, volcanic activity, deep oceanic trenches associated with SUBDUCTION ZONES and emerging fold mountains.

active transport the movement of uncharged molecules against a concentration gradient across a cell membrane or the movement of ions against an electrochemical potential. In both cases, energy is required.

acute an angle that has a size that lies between 0° and 90°.

adaptation a feature of an organism that has evolved under natural selection as it enables the organism to function efficiently in a specific aspect of its particular environmental NICHE. The adaptation can be either psychological, where exposure to certain environmental conditions will only cause a change in the organism's behaviour, or it can be genetic, as when the organism possesses GENES that produce characteristics that prove to be beneficial for survival in its environment. Examples of genetic adaptation are the genetic changes that occur in bacterial populations, conferring antibiotic resistance, as in the strains of bacteria resistant to penicillin. Such antibiotic resistance is a major problem in hospitals when trying to prevent any risk of post-operative infections.

adaptive immune system one functional division of the IMMUNE SYSTEM that produces a specific response to any PATHOGEN. There are two mechanisms of acquired or adaptive immunity:

(1) Humoral—the production of soluble factors, called ANTIBODIES, by activated B-CELLS.

(2) Cell-mediated immunity—the range of T-CELLS involved in the specific recognition of an antigen, which must be present on the surface of another cell, i.e. an antigen-presenting cell (APC). The immunity produced, **adaptive immunity**, is also

11

known as **acquired immunity** because the adaptive immune system is capable of remembering any infectious agent that has induced the proliferation of B-cells (it has acquired a memory for that particular agent, *see* B-CELLS).

adaptive radiation the evolutionary separation of species into numerous descendent species in order to exploit the various habitats that exist throughout the world. This evolutionary divergence of species probably explains the bewildering array of, say, amphibians, which has arisen as a result of adaptive radiation after the first amphibians moved on to land.

addition formula an equation used to express the sum or difference of angles as a sum or difference of the products of the trigonometric functions of the individual angle. For example, the formulae for the functions of cosine and sine are

$$\text{cosine } (A+B) = \cos A \cos B + \sin A \sin B$$
$$\text{sine } (A+B) = \sin A \cos B + \cos A \sin B$$

The above formulae can be used to derive others for functions such as the tangent, $\cos^2 A$, and other trigonometric functions.

addition polymerization the formation of a large molecule called a POLYMER, the structure of which is a repeating array of atoms. The reaction involves the successive addition of molecules of the MONOMER to form a long chain. Addition polymerizations often occur at high temperatures and pressure in the presence of a CATALYST.

addition reaction a chemical reaction in which two parts of one reactant add to a multiple bond of the other reactant, one part to each side of the bond. An addition reaction will usually involve ALKENES, but at the end of the reaction they will no longer contain double bonds but instead have single bonds (the structure of an ALKANE).

adenine ($C_5H_5N_5$) a nitrogenous base component of the NUCLEIC ACIDS, DNA and RNA, which has a PURINE structure. In DNA, adenine will always base pair with THYMINE, but in RNA, during TRANSCRIPTION, adenine will base pair with URACIL. Adenine is also present in other important molecules, such as ATP.

adenosine triphosphate *see* **ATP**.

adiabatic change a change that occurs with no change in heat content, thus if the volume and pressure of an enclosure alter without heat gain from or heat loss to the surroundings, the change is adiabatic.

ADP *see* **ATP**

adrenal gland *see* **endocrine system**.

adsorption the taking up, or concentration of, one substance at the surface of another, e.g. a dissolved substance on the surface of a solid. Chemisorption is when COVALENT bonds hold a layer of atoms or molecules of the dissolved substance on the surface.

aeolian deposits sediments (mainly sand or dust) deposited by the wind (*see* LOESS).

aerobic respiration a set of enzyme-catalysed reactions requiring oxygen to release energy during the degradation of glucose. The first stage of aerobic respiration, GLYCOLYSIS, occurs in the CYTOPLASM of both plant and animal cells. However, the remaining two stages occur in the MITOCHONDRIA of EUCARYOTIC cells and the membrane of PROCARYOTIC cells. Thirty-eight molecules of ATP are generated per molecule of glucose undergoing complete aerobic respiration.

aerosol a fine mist or fog in which the medium of dispersion is a gas. Also, a pressurized container with spray mechanism used extensively for deodorants, insecticides, etc.

agglomerate a rock comprised of essentially angular rock fragments within a fine-grained matrix. A product of explosive volcanic activity.

AIDS (*acronym for* Acquired Immune Deficiency Syndrome) a serious disease thought to be caused by a human retrovirus called HIV human immune deficiency virus which kills the adult T-CELLS of the IMMUNE SYSTEM. This destruction of a critical aspect of the body s immune system leaves the patient open to both minor and major infections, as well as the possibility of developing cancer. Not all HIV-infected individuals develop AIDS, but it can be passed on by the following methods: the receiving of infected blood during transfusions; a mother passing it on to a foetus; the sharing of needles in drug abuse; and the passing of body fluids during sexual contact. There is no evidence that HIV can be transmitted by everyday activities and social contact, such as swimming, sharing cutlery, using a public toilet, etc.

akaryote a cell without a nucleus, or one where the protoplasm of the nucleus does not form a discrete nucleus.

alabaster a fine-grained variety of GYPSUM, often naturally coloured and producing attractive ornaments, etc, when carved.

alcohol an organic compound similar in structure to an ALKANE but with the FUNCTIONAL GROUP -OH (hydroxyl) instead of a hydrogen atom. Alcoholic beverages contain ETHANOL, the alcohol obtained during the fermentation of sugars or starches.

14

aldehyde (*also known as* **alkanal**) any of a class of organic compound with the CO radical attached to both a hydrogen atom and a hydrocarbon (alkyl) radical, giving the type formula R.CO.H.

alga (*plural* **algae**) the common name for a simple water plant, which is without root, stem or leaves but which contains CHLOROPHYLL. Algae range in form from single cells to plants many metres in length. The blue-green algae, cyanobacteria, are widely distributed in many environments.

algebra the use of symbols, particularly letters, in solving mathematical problems in order to find the value of an unknown quantity or to study complex relationships, systems and theories. For example, the physicist, Albert EINSTEIN, used advanced algebra to derive equations from his general theory of RELATIVITY.

algorithm a set of rules comprising a mathematical procedure that enables a problem to be solved in a specific number of steps. Each rule is precise in its definition so that, theoretically, the process can be undertaken by a machine.

aliphatic organic compounds comprising CARBON atoms in OPEN CHAINS (*see* AROMATIC HYDROCARBON). In addition to the main groups, e.g. ALKANES (*see also* open chain), the term includes all derivatives and substitution products (i.e. compounds made by the replacement of an atom or group of atoms in a molecule).

aliquot a small sample of material analysed as being representative of the whole.

alizarin a red solid, $C_{14}H_6O_2(OH)_2$, important for its use as a dye. It is synthetically produced, and used with MORDANTS.

15

alkali a soluble base that will give its solution a pH value greater than 7. The hydroxides of the metallic elements sodium (Na) and potassium (K) are strong alkalis, as is ammonia solution (NH_4OH).

alkali metals the metals lithium (Li), sodium (Na), potassium (K), rubidium (Rb), and caesium (Cs), which belong to group 1A of the PERIODIC TABLE. The last member of the group, francium (Fr), occurs only as a radioactive ISOTOPE. All the metals have one electron in their outer shell and are thus univalent. The elements are all prepared by ELECTROLYSIS of the fused HALIDES. Their melting and boiling points fall with increasing atomic weight, so that caesium has the second lowest melting point of any metal.

alkaline earth metals the metals of group 2A of the PERIODIC TABLE: beryllium (Be), magnesium (Mg), calcium (Ca), strontium (Sr), barium (Ba) and radium (Ra). These bivalent (valency of two) elements have properties similar to the ALKALI METALS but are, in general, less volatile.

alkaloids basic organic substances found in plants and with a ring structure and the general properties of AMINES. In nature, they act as deterrents for the plant against herbivores. Many alkaloids are used in medicine, e.g. cocaine, codeine, MORPHINE, QUININE.

alkanal *see* **aldehyde**.

alkane an open-chain HYDROCARBON with single bonds between each carbon atom. The first member of this HOMOLOGOUS SERIES is METHANE, CH_4, and the subsequent members may be considered to be derived from methane by the simple addition of the unit, $-CH_2$. All the members conform to the general formula of C_nH_{2n+2}, thus the chemical formula for the second member of the series, ETHANE, is C_2H_6. Alkanes are SATURATED COMPOUNDS as all the valence electrons of the carbon atoms are engaged within the single, COVALENT BONDS. As the alkanes are saturated compounds, they are quite stable but will undergo a slow SUBSTITUTION reaction with a HALOGEN. For example:

$$CH_4\,(g) + Cl_2\,(g) \longrightarrow CH_3Cl\,(g)\ +\ HCl\,(g)$$
methane chlorine   chloromethane hydrogen chloride

alkene an open-chain HYDROCARBON containing a carbon-carbon double bond. The first member of this HOMOLOGOUS SERIES is ETHENE, C_2H_4, and all other members conform to the general formula C_nH_{2n}. Alkenes are UNSATURATED hydrocarbons, which can easily undergo an ADDITION REACTION as each carbon atom has one available ELECTRON that is not engaged in the formation of the double bond.

alkyne an open-chain HYDROCARBON containing a carbon-carbon triple bond somewhere in its molecular structure. The first member of the alkynes is ETHYNE (also called acetylene), C_2H_2, and the general formula for the other members is C_nH_{2n-2}. They are UNSATURATED compounds that will readily undergo ADDITION REACTIONS across their triple bond, due to the availability of the four electrons engaged in the formation of two of the carbon-carbon bonds of the triple bond. The remaining carbon-carbon bond of the C=C bond and any carbon-hydrogen bond are a much stronger type of bond, which makes them more stable and less reactive.

allele one of several particular forms of a GENE at a given place (locus) on the CHROMOSOME. Alleles, responsible for certain characteristics of the PHENO-TYPE, are usually present on different chromosomes. It is the dominance of one allele over another (or others) that will determine the phenotype of the individual.

allergy (*plural* **allergies**) an overreactive response of the immune system of the body to foreign ANTI-GENS. Allergies result from the hyperproduction of one class of ANTIBODY, IgE, which activates the release of certain products, including histamine, by the MAST CELLS. This causes the characteristic symptoms of an allergy, i.e. inflammation, itching etc. An induced overreaction by the immune system can be produced by an environmental substance like pollen, certain foods, or even toxins injected by insects such as wasps. Such reactions can be counterattacked using antihistamine drugs.

allotropy the state of an ELEMENT when it exists in two or more forms with different physical properties. Sulphur is the oft-quoted example.

alluvium alluvial deposits i.e. desposited by rivers or streams of recent age. Comprising mud, silt and sand, the sediments are deposited in the river FLOODPLAIN.

Alpha Centauri the brightest of three stars in the constellation Centaurus. Another of the three, Proxima Centauri, is the nearest star to our SUN.

alpha decay the emission of ALPHA PARTICLES from the nucleus of a radioactive ISOTOPE.

alphanumeric a set of characters derived from the numerals 0 to 9 and the alphabet. In computing, the remaining keyboard characters are used for functions other than keying text.

alpha particle a fast-moving helium nucleus ($^4_2He^{2+}$) with a short, straight range from the source of emission.

alpine orogeny the period of mountain building

that occurred for the main part during the Tertiary (*see* APPENDIX 5) and resulted in the formation of the Alps and associated features. The fringe effects of this episode are seen in northern France and the London Basin.

alternating current (AC) an electric current that reverses the direction of its flow at constant intervals of time, resulting in a constant frequency independent of the type of CIRCUIT. Mains electricity is comprised of alternating current as opposed to DIRECT CURRENT.

altitude in geometry, the line segment that measures the perpendicular distance from a vertex of a POLYGON to the opposite side of that polygon, e.g. the height of an isosceles triangle.

altocumulus grey-white sheets, banded layers or rolls of CLOUD irregularly arranged in small segments occurring at medium altitude of 3000–7500 metres (9900–24,500 feet). Sometimes a banded appearance produces a MACKEREL SKY effect.

altostratus greyish sheets or layers of CLOUD, which may appear uniform, fibrous or striated and are composed of water droplets, possibly with ice crystals. Occurring at 3000–7500 metres (9900–24,500 feet), it often thins sufficiently to allow sunshine through.

aluminium a light, ductile and malleable metal, which is a good conductor of electricity. It is found in group 3A of the PERIODIC TABLE and is the most common metallic element in the earth's crust (third most common overall, at 8 per cent). It is extracted from the hydrated oxide (bauxite) by electrolysis with molten cryolite as a FLUX. The metal and its alloys are used for innumerable purposes, including the manufacture of cooking utensils, electrical equipment, aircraft, etc.

amalgam the alloy of a metal with MERCURY.

American Standard Code for Information Interchange *see* **ASCII**.

amides organic compounds where the hydroxyl (-OH) from the carboxyl group (-COOH) in acids has been replaced by the amide group -NH_2. In effect they are ammonia derivatives with the general formula $RCONH_2$.

amines compounds formed from ammonia, NH_3, by the replacement of one or more hydrogen atoms with organic RADICALS. There are three classes: primary, NH_2R; secondary, NHR_2, and tertiary, NR_3.

amino acid any of the 20 standard organic compounds that serve as the building blocks of all PROTEINS. All have the same basic structure, containing an acidic carboxyl group (-COOH) and an amino group (-NH_2), both bonded to the same central carbon atom referred to as the α-carbon. Their different chemical and physical properties result from one variable group, the side chain or R-group, which is also attached to the α-carbon.

amino group an essential part (-NH_2) of the AMINO ACIDS.

ammonia a COVALENT compound (NH_3) that exists as a colourless gas. It will react with water, giving the alkaline solution known as ammonium hydroxide (NH_4OH). Ammonia will ionize in water to form the ammonium ion (NH_4^+) and the hydroxide ion (OH^-)
$$NH_3 \text{ (g)} + H_2O \text{ (l)} \longrightarrow NH_4^+\text{(aq)} + OH^-\text{(aq)}.$$

ammonite a fossil subclass of cephalopods which were abundant from the Devonian to the Upper Cretaceous (*see* APPENDIX 5). All species are now extinct. The shell, which was commonly coiled, contained chambers and often exhibited outer marks and patterns.

amorphous having no shape or form—non-crystalline.

ampere (A) the unit for measuring the quantity of electric CURRENT, often abbreviated to **AMPS**.

amphiboles an important group of rock-forming minerals comprising silicate structures with iron, calcium, magnesium, sodium and aluminium. They are found in many igneous and metamorphic rock types, and hornblende, $NaCa_2(Mg,Fe,Al)_5 (Al,Si)_8 O_{22} (OH)_2$, is a commonly occurring member of the group.

amphoterism a property of the oxides or hydroxides of certain metallic elements, which allows them to function both as acids and alkalis. Although insoluble in water, amphoteric compounds will dissolve in acidic solutions (pH < 7) or basic solutions (pH > 7).

amplitude the maximum displacement of a particle from its position of rest. For sound waves, the amplitude relates to the intensity of the wave and is the distance between the X-AXIS (rest position) and the crest or trough of the wave. In a simple pendulum, the amplitude is defined as the angle through which the arm moves when swinging between its extreme and rest positions.

AMPS *see* **ampere**.

anabolism the biosynthetic building-up processes of METABOLISM. For example, the synthesis of fatty acids and PHOSPHOLIPIDS in the GOLGI APPARATUS.

anaemia *see* **erythrocyte**.

anaerobic respiration a set of enzyme-catalysed reactions releasing energy from the SUBSTRATE in the absence of oxygen. The first stage of anaerobic respiration is GLYCOLYSIS—as in AEROBIC RESPIRATION—but there is only one other stage, FERMENTATION, which has two alternative pathways as follows:

(1) The production of lactic acid, as in muscle cells during prolonged contraction.

(2) The production of the alcohol, ethanol, by micro-organisms such as yeast or some plants.

In all anaerobic respiration, only two ATP molecules are generated (during glycolysis) per glucose molecule.

anemometer a device that measures the speed of the wind and often comprises 4 cups at the end of arms—the rotating cups anemometer. There is also a pressure-driven variety that relies upon the wind pressure through a tube (pressure-tube anemometer).

angström the unit of measurement (10^{-10}m), now superseded by the nanometre ($10\text{Å} = 1$nm), which was used for electromagnetic radiation. Also used by crystallographers for interatomic measurements.

angular momentum the momentum of a body rotating around a point. For a body in motion through an angle about an axis (the angular velocity), the angular momentum is the product of the angular velocity and the moment of inertia (the mass of the body x the square of its distance from the axis). The Earth has both rotational angular momentum (rotating on its axis) and orbital angular momentum (orbiting the SUN).

anhydrite an EVAPORITE mineral, $CaSO_4$, which occurs in a soft, whitish, usually AMORPHOUS form. It readily forms gypsum and may occur as a CAP ROCK above SALT DOMES. It is used in the manufacture of plaster and cement.

animal starch *see* **glycogen**.

anion a negatively charged ion formed by an atom or group of atoms gaining one or more electrons.

anisotropy the state of a substance when it possesses a property dependent upon direction, e.g. some crystals have a different REFRACTIVE INDEX in different directions. The opposite of ISOTROPY.

annealing the process whereby materials, usually metals, are heated to, and held at, a specific temperature before controlled cooling. This relieves STRAIN set up by other processes.

annual parallax as the EARTH moves around the SUN the motion causes very small positional changes to the stars. The annual displacement is termed the annual parallax.

annulus the shape created by the area between two concentric circles, as with a washer.

anode the positively charged electrode of an electrochemical cell to which the ANIONS of the solution move to give up their extra electrons, i.e. the electrode at which OXIDATION occurs.

anodizing the process whereby an oxide coat is deposited on the surface of a metal (often aluminium or an alloy) by making the metal the anode in ELECTROLYSIS. The oxide skin forms a protective layer and may be rendered decorative through the use of a dye in the electrolytic process.

antibiotic a chemical produced by micro-organisms, such as BACTERIA and moulds, that can kill bacteria or prevent their growth. The first antibiotic to be discovered was penicillin, and there are now many more, including erythromycin, streptomycin and terramycin.

antibody (*plural* **antibodies**) a protein circulating in the blood, which is produced by B-CELLS and will

23

bind to the surface of an ANTIGEN. The production of antibodies is a specific immune response of the ADAPTIVE IMMUNE SYSTEM. Antibodies consist of protein chains that form IMMUNOGLOBIN, i.e. Ig, and are very useful for identifying specific types of protein unique to a particular plant or animal or to a VIRUS that may be circulating throughout the body of an individual. Although millions of different antibodies are produced in order to cope with any PATHOGEN that may arise, there are only five major classes, which have the following functions:

Antibody	Function
IgG	The most abundant, it combats microorganisms and toxins. As it can cross the placenta, it is the first Ig found in newly born infants.
IgA	The major Ig in mucosal secretion, it defends external surfaces including the gut wall to help cope with antigens found within the gut.
IgM	The first Ig to be produced during infection, it is very effective against bacterial infections.
IgD	This is present on surfaces of B-cells, but no specific function is known for it.
IgE	This protects external surfaces and triggers the release of histamine from other cells of the immune system.

anticline a FOLD in rock strata where the strata (fold limbs) dip away from the fold axis. Under normal conditions an anticline contains beds that become younger upwards.

anticyclone an area of pressure that increases to and reaches its highest at the centre. Winds tend to be light, reflecting small pressure gradients and flow clockwise in the northern hemisphere and anticlockwise in the southern. Associated weather tends to be settled and fine, although anticyclones in winter can give rise to cold conditions and often FOG.

antigen any substance that triggers an immune response due to the body s IMMUNE SYSTEM recognizing it as foreign. Common antigens are proteins present on the surface of bacteria and viruses. Unsuccessful transplant operations are usually a result of the patient s immune response recognizing the surface cells of the organ from the donor as non-self. The organ is said to be rejected when the patient's immune system becomes activated and tries to destroy the donated organ.

antimatter matter, as yet hypothetical, that is composed of **antiparticles**, i.e. particles of the same mass but opposite values for its other properties such as charge. The ELECTRON and POSITRON are particle and antiparticle, and interaction between them would result in annihilation and the release of energy.

antioxidant a substance added to some materials, such as paint, oils and rubber, to delay the harmful process of oxidation.

antiparticles *see* **antimatter**.

aorta the largest ARTERY in the body, which acts as the blood outflow of the left VENTRICLE of the heart. The aorta has an approximate diameter of three centimetres, with thick muscular walls to carry blood under pressure. The aorta divides into several branches, which supply blood to the arms and the head. It then continues down around the spine to the level of the lower abdomen, where it divides into two major branches to supply the legs.

apatite a calcium phosphate mineral, also containing fluorine, chlorine, and HYDROXYL ions. It occurs as an accessory mineral in IGNEOUS and METAMORPHIC rocks, and is the primary constituent of fossil bones and vertebrate teeth. It is used industrially in the manufacture of fertilizers.

APC *abbreviation for* antigen-presenting cell. *See* ADAPTIVE IMMUNE SYSTEM.

apogee the point at which a satellite is at its greatest distance from the earth. The converse situation is called the perigee.

aquifer a permeable ROCK, underlain by impermeable strata, which contains significant quantities of recoverable water.

Archimedes' principle a law of physics stating that when a body is partly or totally immersed in a liquid, the apparent loss in weight equals the weight of the displaced liquid.

arenaceous rocks SEDIMENTARY rocks in which sand grains form a major component. Three main groups are defined—quartz sandstones, which have 95 per cent QUARTZ; ARKOSES, which have > 25 per cent FELDSPAR; and GREYWACKES, which are poorly sorted sediments.

argillaceous rocks sediments that are composed of silt/clay-sized particles. More than half of SEDIMENTARY rocks are argillaceous, and most have a high proportion of CLAY minerals. Organic material is often present, providing potential as source rocks for HYDROCARBONS.

argon one of the INERT gases. It is unreactive and is used in lamps, fluorescent tubes, etc.

arithmetic mean an average. For a set of numbers with n values, the mean is the sum of the numbers divided by n.

arithmetic series the sum of the terms of a sequence of quantities. An unknown term (nth term) can be calculated using the formula below:

$$n^{th} \text{ term} = a + (n - 1)d$$

The sum of n terms is calculated using:

$$S_n = n/2 \, [2a + (n - 1)d]$$

where "a" is the first term and "d" is the common difference of the series.

arkose a sandstone containing quartz and ≥ 25 per cent feldspar. Evidence suggests arkoses were rapidly deposited under acid conditions, probably near to land and an eroding GRANITE/GREISS source.

armature the rotating coil of an electric motor. More generally, it is any electric component within a piece of equipment in which a VOLTAGE is induced by a MAGNETIC FIELD.

aromatic hydrocarbon any compound that has a molecular structure based on that of BENZENE. Aromatic hydrocarbons are UNSATURATED, closed-chain compounds that will undergo SUBSTITUTION reactions as well as ADDITION REACTIONS, depending on which functional groups are present within their structure and the reactivity of the other reagents.

arteriosclerosis the thickening, hardening and loss of elasticity of the ARTERIES. This can be a pathological condition of advancing age, or it can be associated with fatty deposits, particularly CHOLESTEROL, that block the arteries, causing their diameter to decrease. As a result, the heart must strain to increase its muscular activity to generate enough pressure to pump the blood through the arteries.

artery (*plural* **arteries**) a thick-walled vessel that carries blood under pressure resulting from the pumping mechanism of the heart. The PULMONARY ARTERY carries deoxygenated blood from the heart to the lungs, but all other arteries carry oxygenated blood from the heart to body tissues.

artificial intelligence the concept that computers can be developed to simulate human intelligence, including learning, reasoning, and adaptation. Also, a branch of computer science including *inter alia* robotics, pattern recognition, and knowledge-based systems.

asbestos fibrous silicate minerals within the AMPHIBOLE and serpentine groups. The principal varieties are chrysotile (serpentine group) and crocidolite, actinolite, and others from the amphibole group. The minerals resist heat and are used in the manufacture of fire-resistant fabrics, brake linings, etc. The fibres are toxic if inhaled.

ASCII (*acronym for* American Standard Code for Information Interchange) a standard code of 128 ALPHANUMERIC characters for storing and exchanging information between computer programs.

ascorbic acid another term for the water-soluble vitamin C found in all citrus fruits and green vegetables (especially peppers). A deficiency of ascorbic acid leads to the fragility of tendons, blood vessels and skin, all of which are characteristic of the disease called scurvy. The prescence of ascorbic acid is also believed to help in the uptake of iron during the process of digestion by the body.

asphalt a naturally occurring black/brown semi-solid substance containing bitumen and mineral matter. It occurs in oil-bearing strata after removal of volatiles and can form "pools", e.g. in California, or "lakes", as in Trinidad. In civil engineering, it is mixed with chips of SANDSTONE, LIMESTONE or GRANITE for road laying, and other construction materials.

assay the chemical analysis of a mixture to determine the amount of a particular constituent, e.g. metal in an ore.

asteroid one of many rocky or metallic bodies, more correctly called planetoids, that orbit the sun between the orbits of Mars and Jupiter. Most are very small, but the largest, Ceres, has a diameter of about 1,000 kilometres. Meteorites are debris from the asteroid belt formed by the collision of the bodies.

asthenosphere the higher layer of the upper mantle of the EARTH, which, in comparison to the LITHOSPHERE upon which it rests, is relatively weak. The zone that accommodates lateral, plastic flow is partly molten and is the site for MAGMA generation.

astronomical unit (AU) the average distance from the centre of the EARTH to the centre of the SUN and equal to 1.496×10^8km (92.9×10^6 miles). The AU is used as a measure of distance within the solar system.

asymptote a straight line that a curve of a FUNCTION approaches but never reaches. Although the perpendicular distance between the curve and its asymptote will decrease and eventually equal zero, the distance between the origin and the asymptote will tend to infinity, and there will still be no contact between the curve and the asymptote.

atmosphere (1) a unit of PRESSURE defined as the pressure that will support a mercury column 760mm high at 0°C, sea level, and a latitude of 45°. (2) the

29

layer of gases surrounding the earth, which contains, on average, 78 per cent nitrogen, 21 per cent oxygen, almost 1 per cent argon, and then very small quantities of carbon dioxide, neon, helium, krypton and xenon. In addition, air usually contains water vapour, HYDROCARBONS and traces of other materials and compounds.

atom the smallest particle that makes up all matter and still retains the chemical properties of the element. Atoms consist of a minute nucleus containing PROTONS (p) and NEUTRONS (n), with negatively charged particles called ELECTRONS (e) moving around the nucleus. The number of protons in an atom is equal to the number of electrons, and as the protons are positively charged, the atom is, overall, electrically neutral (*compare* ISOTOPE). The various elements of the PERIODIC TABLE all have a unique number of protons within their atomic nucleus.

atom bomb *see* **nuclear fission**.

atomic mass unit (*see* RELATIVE ATOMIC MASS) defined in 1961 as one twelfth of the mass of an atom of ^{12}C, having formerly been based upon ^{16}O, the most abundant ISOTOPE of oxygen.

atomic number (A, at. no.) the number of protons in the NUCLEUS of an ATOM. Although all atoms of the same element will have the same number of protons, they can differ in their number of neutrons, resulting in an ISOTOPE.

atomic scattering the process whereby radiation (often X-rays or electrons) is deflected upon entering a medium or structure due to interaction/collision with the atoms of the material.

atomic weight *see* **relative atomic mass**.

ATP (*abbreviation for* adenosine triphosphate) an

important molecule that is used as energy to drive all cellular processes. It consists of ADENINE and a 5-ring sugar that has three phosphate (PO_4) groups attached by high-energy bonds. ATP can be synthesized during GLYCOLYSIS by addition of a phosphate group to adenosine diphosphate (ADP), or it can be broken down to form the ADP, releasing energy that will be used to drive a metabolic process, such as active transport across cell membranes or the contraction of muscle cells.

atrium (*plural* **atria**) a minor chamber of the heart that is considered to be a reservoir as blood passes from it into the pumping chamber, the VENTRICLE. The right atrium of the heart receives the blood carried by the superior and inferior VENA CAVA before it passes via a valve into the right ventricle. The pulmonary veins carry oxygenated blood from the lungs into the left atrium, which then flows via a valve into the left ventricle.

aureole in geology, the contact or metamorphic aureole is the area around an igneous INTRUSION where the host rocks have been subjected to heat, producing recrystallization, i.e. thermal metamorphism. In meteorology, a ring (the inner part of the CORONA) sometimes seen around the Sun or Moon. The effect is created by DIFFRACTION of light by water droplets in high cloud formations.

aurora luminous, and often colourful, sheets or streaks in the sky, formed by high-speed, electrically charged solar particles entering the upper atmosphere where electrons are released, thus creating molecules with the associated release of light. These effects are related to SUNSPOT activity and are termed the Northern Lights (*aurora borealis*) in the northern hemisphere, and the Southern Lights (*aurora australis*) in the southern hemisphere.

autolysis *see* **lysis**.

autosome a biological term describing all CHROMOSOMES within a cell except the SEX CHROMOSOMES. In a DIPLOID cell, there are two copies of every autosome, each of which will carry genetic information for the same aspect of the individual's PHENOTYPE. Although autosomes are not involved with determining the sex of an individual, they can carry genetic information that will affect the sexual characteristics.

Avogadro's constant *or* **Avogadro's number** the number of particles present in one MOLE of a substance. It is given the symbol N or L and has the value of 6.023×10^{23}.

Avogadro's law the principle formulated by the Italian scientist Amedeo Avogadro (1776-1856) that states that equal volumes of all gases contain the same number of molecules when under the same temperature and pressure. For the purpose of calculation, one MOLE of gas will occupy a volume of 22.4 litres at standard conditions, i.e. 273.15K and 1 atmosphere.

axon *see* **neuron**.

azeotropic mixtures a mixture with a constant boiling point and a constant composition despite generation and removal of vapour on boiling. The boiling point of the mixture can be higher or lower than the boiling points of the components.

azimuth the angle between the vertical plane containing a line or body and the plane of the MERIDIAN.

B

bacillus (*plural* **bacilli**) a general term for any rod-
shaped bacterium. Also a genus of bacterium found
in soil and the air.

backcross a mating experiment that is used to
discover the GENOTYPE of an organism. It involves
crossing the organism of unknown genotype with an
organism of known genotype (usually the HOMOZYGOTE
recessive). The PHENOTYPES of the produced progeny
should directly correspond to the chromosomes of
the parental organism of unknown genotype. A
backcross usually reveals whether the unknown
genotype is homozygous or heterozygous (*see*
HETEROZYGOTE) for a particular gene.

background when the measurement of any signal is
affected by additional signals from sources other
than the one being measured. These incidental
values must be accounted for when calculating
readings. The effect is due to natural radioactivity,
COSMIC RAYS, etc.

back scatter when a surface is subjected to radiation or a stream of particles, some of the incident
beam is reflected back towards the source. This is
back scatter.

bacteriophage *or* **phage** a virus that infects a
BACTERIUM. Being specific to a particular bacterium,

phages find use in genetic engineering as a vehicle in cloning (*see* GENETIC ENGINEERING) and in certain manufacturing processes that utilise bacteria, e.g. the production of cheese.

bacterium (*plural* **bacteria**) a micro-organism, usually unicellular, which does not photosynthesize (*see* PHOTOSYNTHESIS) and which causes diseases that are treated with ANTIBIOTICS. Bacteria occur in water, air, soil and rotting animal or plant debris (SAPROTROPHS). They are classified by shape into three main groups: the spherical coccus form; the spiral spirillum; and the rod-shaped bacillus.

ballistics the study of the flight path of an object moving under the influence of gravity.

bar chart *or* **bar graph** a graph that illustrates relationships between variables by vertical parallel bars, with the heights of the bars representing the data proportionally.

barometer an instrument used to measure the pressure that the atmosphere exerts on the earth s surface, which helps to predict impending weather changes.

barrier reef a reef of coral built up parallel to the shore but some way from it, so as to create a lagoon. A good example is the Great Barrier Reef, Australia, which is almost 2000 km (1243 miles) long.

Barrovian metamorphism *or* **Barrow's zones** (*see also* METAMORPHIC ROCKS, INDEX MINERALS) a sequence of metamorphic mineral assemblages created by the regional metamorphism of clay-rich sedimentary rocks. In the original study in northeast Scotland, George Barrow identified a number of zones, each with a different, characteristic mineral indicative of increasing metamorphic grade. In

PELITIC ROCKS, chlorite is the low-grade mineral, and with increasing grade BIOTITE (mica), GARNET, kyanite and finally sillimanite develop.

basalt a dark fine-grained IGNEOUS ROCK containing FELDSPAR (plagioclase), PYROXENE and possibly OLIVINE. Basalt occurs as LAVAS (extrusion) and minor INTRUSIONS. Basalt flows cover a vast proportion of the Earth's surface (over two-thirds) and the "terrestrial" planets (MERCURY, VENUS and MARS). Two subdivisions exist—alkali basalts and tholeiites—the former being found in regions of rifting, crustal deformation and on oceanic islands, while the latter typify ocean floor and stable continental crust extrusion.

base in a chemical reaction, any substance that dissociates in water to produce hydroxide ions (OH⁻). A base can be thought of as a proton (H⁺) acceptor or as an electron pair donor. It will react with an acid to give a salt and water (the latter formed from the OH⁻ ion from the base and the H⁺ ion from the acid).

$$NaOH(aq) + HCl(aq) \longrightarrow NaCl(aq) + H_2O$$

In mathematics, a base is the number raised to a certain power (EXPONENT), which will produce a fixed number. For example, base 5 to the power of 3 will equal 125, i.e. $5^3 = 125$.

basement igneous or metamorphic rocks, often highly deformed, over which sediments lie unconformably (*see* UNCONFORMITY). Basement often equates with Precambrian rocks, i.e. the oldest rocks, formed in the period ending 590 million years ago (*see* APPENDIX 5).

base pair the arrangement of two nitrogenous molecules on opposite strands of a DNA double-helix or a DNA-RNA molecule. The bases can be classified according to their structure: ADENINE and GUANINE are PURINES, whereas THYMINE, CYTOSINE and URACIL are PYRIMIDINES. As a consequence of geometrical factors, a purine will be hydrogen-bonded to a pyrimidine, i.e. A-T and C-G. This specific base-pairing keeps the DNA structure in a highly organized order, allowing the replication of DNA to be very precise, thus ensuring that each daughter cell will inherit the same genetic information contained within the parent cell.

BASIC (*acronym for* Beginners All-purpose Symbolic Instruction Code) a simple language for procedural programming and interacting with a computer.

basic rocks IGNEOUS rocks, e.g. gabbro and BASALT, with a low SILICA content (45–53 per cent by weight) and a high proportion of magnesium, iron and calcium.

batholith a very large igneous INTRUSION, often of granitic composition. Such bodies are usually elongate, may be hundreds of kilometres long, and often comprise several plutons (intrusive igneous body) that merge at depth. Batholiths are discordant, i.e. they cut across the country (host) rocks.

bauxite the primary ore source for ALUMINIUM, formed by the weathering of aluminium-bearing rocks in tropical conditions. A residual deposit that, if it contains > 25 per cent aluminium oxide, can be exploited commercially.

B-cells these are LYMPHOCYTES, which differentiate in the bone marrow to form part of the IMMUNE SYSTEM of humans. B-cells become activated when they encounter a specific ANTIGEN, leading to proliferation and secretion of ANTIBODIES by the activated B-cells. After a first encounter with an antigen, some of the B-cells remain and serve as memory cells. The memory cells will be capable of recognizing the same antigen during any subsequent encounter and will, therefore, produce a faster and greater secondary response (this is the principle behind vaccination).

Beaufort scale a system for indicating wind velocity, with measurements taken at 10 metres (32.8 ft) above ground level. The numerical scale ranges from 0 (calm, speed < 0.3 ms^{-1}) to 12 (hurricane, speed > 32.7 ms^{-1}).

becquerel the unit of radioactivity in the SI scheme, defined as the activity of a radionuclide decaying at one nuclear transition per second. It is named after Antoine Henri Becquerel (1852–1908) who began the study of radioactivity.

bedding plane the plane surface that separates beds in SEDIMENTARY rocks. Each plane reflects a break in deposition and may be caused by differences in composition, grain size, etc.

benthic a plant or animal living on the bottom of a lake or sea. This may include crawling or burrowing at the sediment-water boundary, being attached to it (as with seaweeds), or purely sessile.

bentonite a useful clay formed by the breakdown of volcanic ash, which has properties similar to FULLER'S EARTH. It is THIXOTROPIC and is used in the construction, paper and pharmaceutical industries and as an additive in oil-drilling muds.

benzene a toxic hydrocarbon that is a liquid at room temperature and has the chemical formula C_6H_6. The six carbon atoms of benzene form a ring, and the overall molecular structure is planar, with all bond angles having the same value of 120°. The benzene ring is a very stable structure due to the delocalization of six electrons (one electron is contributed by each carbon atom) and therefore does not readily undergo an ADDITION REACTION. All ARO-MATIC HYDROCARBONS contain a benzene ring within their molecular structures.

beri-beri a crippling human disease that is caused by dietary deficiency of the water-soluble vitamin B, also called thiamine, which is essential for the metabolic conversion of carbohydrate to glucose. Affected individuals show symptoms of muscle atrophy, paralysis, and mental confusion, and may eventually suffer from heart failure. Chronic alcoholism will lead to beri-beri, and this is thought to be responsible for the development of a psychosis called KORSAKOFF'S SYNDROME.

beryl naturally occurring beryllium alumino-silicate, $Be_3Al_2Si_6O_{18}$. It forms an accessory mineral in some igneous and metamorphic rocks and also more extensively in PEGMATITES. Coloured varieties are precious stones—emerald (green), aquamarine (blue-green), and morganite (pink). The metal beryllium is extracted from beryl by ELECTROLYSIS.

beta decay a type of ionizing radiation that emits either a negatively charged ß-particle (e⁻, an electron) or a positively charged ß-particle (e⁺, a positron). In electron emission, the PROTON number of the nucleus increases by one, whereas in positron emission, the proton number decreases by one. Both types of ß-particles tend to have a longer range than those emitted during ALPHA DECAY and produce a scattered rather than a straight path through matter.

bicarbonates acid salts of carbonic acid (H_2CO_3), where one of the hydrogens has been replaced by a metal. Aqueous solutions contain the bicarbonate ion (HCO_3^-).

bilateral symmetry the case when an organism can be "split" into two halves that are nearly mirror images of each other.

bile a viscous fluid produced by the liver and stored in a small organ, the gall bladder, near the liver. Bile is an alkaline solution consisting of bile salts, bile pigments and CHOLESTEROL, which aids in the digestion of fatty particles present in the diet. Food entering the DUODENUM triggers the muscular contraction of the gall bladder wall, and bile is forced into the duodenum via the bile duct. Although bile does not contain any digestive enzymes, the bile salts help in the digestion of fatty food particles by acting as emulsifying agents, i.e. they break down large fat particles into many smaller ones, a process that exposes a larger surface area to the digestive action of the enzymes, LIPASES.

binary code *see* BINARY SYSTEM

binary digit *see* **bit**.

binary star two stars that are mutually attracted by gravitational forces, thus forming a double star where both bodies revolve around a common centre of mass.

binary system an arithmetical code that uses a combination of the two digits, 1 and 0, expressed to base 2. The value of any digit in a binary number increases by powers of 2 with each move to the left. For example, 1010.1 to base 2 in the binary system (written 1010.1^2) represents:

$$(1 \times 2^3) + (0 \times 2^2) + (1 \times 2^1) + (0 \times 2^0) + (1 \times 2^{-1})$$

which, in the everyday decimal system, adds up to:

$$8 + 0 + 2 + 0 + \frac{1}{2} = 10\frac{1}{2}.$$

The binary number system is the basis of the internal coding in all modern computing, as the two digits, 0 and 1, are represented as the on/off states of any switch in a circuit.

binomial theorem a mathematical formula whereby the expansion of any positive power of a binomial (a + b) to a POLYNOMIAL may be reached without performing the numerous multiplications:

$$(a + b)^n = a^n + n a^{n-1} b + n(n - 1)a^{n-2}b^2/2! + \ldots + b^n.$$

For example, the binomial $(a + b)^5$ has the following expansion:

$$(a + b)^5 = a^5 + 5a^4b + 10a^3b^2 + 10a^2b^3 + 5ab^4 + b^5$$

binomics *see* **ecology**.

biochemistry the investigative study of chemical aspects of the biological processes occurring in the cells of all organisms. Such investigations provide an understanding of a broad range of important processes, from the control of cell metabolism to the effects that a certain disease has upon the cells of the body.

biological (*or* **biochemical**) **oxygen demand** (BOD) a measure of the pollution of effluent where micro-organisms take up dissolved oxygen in decomposing the organic material present in the effluent. BOD is quantified as the amount of oxygen, in milligrams, used by one litre of sample that is stored in the dark at 20°C for five days.

biosphere the region of the earth's surface (both land and water) and its immediate atmosphere, which can support any living organism.

biosynthesis the production of complex chemical compounds by living organisms using ENZYMES.

biotechnology the industrial use of organisms, their parts or processes, to produce drugs, food or other useful products. Modern processes include the controlled growth of specific fungi in laboratories to obtain the antibiotic, penicillin, and the production of alcohol during FERMENTATION in yeast. The scope for biotechnology is enormous, with a great deal of research being directed towards GENETIC ENGINEERING in plants and animals.

biotic relating to life or living things (hence biota, the animal and plant life of a region.) Thus for an organism, the other living organisms around it comprise the biotic environment and may be competitors, predators, parasites, etc.

biotite a dark-coloured MICA and widespread rock-forming mineral occurring in IGNEOUS (especially granites), METAMORPHIC (schists, gneisses, etc) and also some SEDIMENTARY rocks. Compositionally it is a complex silicate of iron, magnesiuum and aluminium with potassium.

birefringence the double REFRACTION of light through a crystal, whereby plane-polarized light is split into two rays vibrating in planes at right angles to each other and travelling along different path lengths through the mineral, producing a retardation. The result is that two images are created.

In the study of minerals with a microscope, the retardation described produces a range of interference colours when the mineral is viewed through crossed polarized light.

bit (*abbreviation for* **binary digit**) either the number 1 or 0. A bit is the smallest unit of a computer capable of storing one unit of information, although the storage capacity of the computer is measured in BYTES.

black body a body that absorbs all heat or light radiation falling upon it. In practice, no body can achieve this state.

black hole a region in space from which matter and radiation cannot escape due to the intense gravitational field. Their origin is thought to lie with the collapse of massive stars. Since no radiation is emitted, black holes cannot be observed directly. However, it is believed that visible stars form BINARY STARS with black holes and the capture of matter by the black hole from the visible star allows indirect observation, because X-rays are radiated as the matter falls into the black hole. (*See also* EVENT HORIZON).

blood a vital substance consisting of red blood cells (ERYTHROCYTES) and white blood cells (LEUCOCYTES) suspended in a liquid medium called BLOOD PLASMA. As well as many proteins, mammalian blood also contains small disc-shaped cells called platelets, which are involved in BLOOD CLOTTING. Blood circulates throughout the body and serves as a mechanism for transporting many substances. Some of the essential functions of blood are:

(1) Oxygenated blood is carried from the heart to all tissues by the arteries while the veins carry deoxygenated blood, which contains carbon dioxide, back to the heart.

(2) Essential nutrients, such as glucose, fats and AMINO ACIDS (the building blocks of proteins), enter the blood from the intestinal wall, or the liver and fatty deposits, and are carried to all the regions of the body.

(3) The metabolic waste products, ammonia and carbon dioxide, are carried to the liver, where they

react to form urea, which is then carried by the blood to the kidneys for excretion.

(4) Steroid and thyroid hormones important regulatory molecules are carried to their target cells after they are secreted into the blood by the ENDOCRINE SYSTEM. Although insoluble molecules, they are carried in soluble particles called low density LIPOPROTEINS.

blood clotting (*also called* **haemostasis**) a process, involving many chemical factors, that stops blood leaking from an area of injured tissue. In the first instance, constriction of any blood vessels in the injured tissue restricts the leakage of blood, and the subsequent formation of a plug helps to seal off the damaged area, preventing the entry of micro-organisms. The formation of this plug is triggered by an enzyme secreted by damaged blood vessels and blood platelets and is completed after a chain of chemical reactions. The final hard clot consists of blood platelets, trapped red blood cells, and fibrin (*see* FIBRINOGEN).

blood grouping a method for classifying blood types by checking which particular ANTIGENS are present on the surface of red blood cells. More than four hundred antigens can be recognized by their specificity for a particular ANTIBODY. There are many systems for the classification of blood types, but the ABO blood group and rhesus (Rh) group are two important systems that are widely known. In the ABO blood group system, there are basically two antigens, designated A and B. The A and B antigen may be present singly or together (AB), and the absence of both antigens gives rise to blood group O. There are thus four blood groups A, B, AB or O.

Naturally occurring antibodies to the ABO system develop only after the age of three months. A person of blood group A will have antibody-B present in their serum, i.e. anti-B serum. A person of blood group B has anti-A in their serum. A person of blood group O has both anti-A and anti-B, whilst a person of blood group AB has no antibodies.

In blood transfusions, it is vital that blood groups are correctly matched since incompatibility will result in blood clotting, which could cause the recipient s death. For instance, if the donor is blood group A and the recipient is blood group B, then anti-B serum of the donor will react with the B-antigen present in the recipient s blood and initiate blood clotting. Of course, the A-antigen of the donor's blood will also react with the anti-A in the recipient's blood.

The rhesus blood group system can be simply explained in terms of whether an individual is Rh-positive or Rh-negative. The presence of an Rh-factor (D-antigen) on the surface of red blood cells will classify an individual as Rh-positive, whereas the absence of such a factor will classify the individual as Rh-negative.

blood plasma blood from which all the blood cells (ERYTHROCYTES, LEUCOCYTES and platelets) have been removed. The resulting solution is 90 per cent water and contains some proteins, sugar, salt, urea, hormones and vitamins.

blood serum a fraction of the liquid medium of blood, i.e. plasma, minus one of the plasma proteins, called FIBRINOGEN.

blood type *see* **blood grouping**.

BOD *see* **biological oxygen demand**.

Bohr, Niels Henrik David (1885-1962) a Danish physicist who carried out crucial research on the QUANTUM theory. In 1922 he received a Nobel prize, and in 1952 he helped establish CERN, the European organization for nuclear research, which has a laboratory in Geneva for the investigation of the physical properties of high-energy particles. He created the theory of an electron-proton atom, and his son, the physicist Aage Bohr (1922-), was part of the Nobel prize-winning team of 1975 for research on the atomic nucleus theory.

Bohr effect the discovery by the Danish physiologist Christian Bohr (1855-1911) that the oxygen-carrying capacity of blood varies within different parts of the body. He showed that the pH of the body tissue determined whether oxygen would be released from, or taken up by, the HAEMOGLOBIN in red blood cells. At low pH (acid conditions), oxygen is released from haemoglobin and enters the surrounding tissues, but at high pH (alkali conditions), oxygen from the surrounding tissue will bind strongly to the haemoglobin in the red blood cells. The Bohr effect explains why oxygen is taken up by the haemoglobin in blood circulating throughout lung tissue (high pH) but is released by blood circulating in active muscle sites (low pH).

Bohr theory a theory proposed by Niels BOHR to explain the structure of the atom. In essence, he postulated that an electron in an atom moves in circular orbits about the nucleus, in a so-called stationary state of energy. When electrons move between orbits, energy is absorbed or emitted, in the latter case as light. The theory has now been superseded by the concept of wave mechanics, which deals more adequately with complexities introduced by atoms with two or more electrons.

boiling point the temperature at which a substance changes from the liquid state to the gaseous state. It occurs when the vapour pressure of a liquid surface equals the surrounding atmospheric pressure. The boiling point of a pure liquid is measured under the standard atmospheric pressure of 1 atmosphere (equivalent to 760mm mercury).

bond the force that holds atoms together to form a MOLECULE, a compound or a lattice. The type of bond between neighbouring atoms will be determined by the electron attraction strength of each atom and can be of three types COVALENT, IONIC or POLAR COVALENT.

bond dissociation energy the energy required to break a specific bond that holds two atoms together in a diatomic molecule. For example, 949 kJ are needed to carry out the dissociation of a diatomic nitrogen molecule (N_2) into two separate atoms.

bond length the distance between atoms in a covalently bonded compound. Some typical lengths are carbon-hydrogen 107 picometres; oxygen-hydrogen 96 picometres; carbon-carbon (single bond) 154 picometres; and carbon-carbon (double bond) 133 picometres.

botryoidal a term meaning "resembling a bunch of grapes in shape," used in mineralogy to describe mineral HABIT.

botulism the most dangerous type of food poisoning in the world, caused by the anaerobic bacterium called *Clostridium* botulism. This bacterium is found in an oxygen-free environment, such as underneath soil or in an airtight food can. During growth, it releases toxins that, if bacteria are living in the cells of the body, will affect the nervous system of humans. This can result in death, especially if it is the vulnerable members of a population who are affected, e.g. newborn babies and elderly people.

boulder clay (*see also* TILL) a deposit of glacial origin formed of boulders in an ill-sorted matrix of mainly clay. Being laid down beneath ice leaves an unstratified deposit containing rocks from the terrain traversed by the glacier or ice sheet.

Boyle's law the principle that a volume of gas varies inversely with the pressure of the gas when the temperature is constant. Thus, under constant temperature, the doubling of the pressure will lead to the volume being halved (*see also* GAS LAWS). The British scientist Robert Boyle (1627-1691) not only enunciated this law but he was also one of the original founders of the oldest British scientific society, the Royal Society.

bp *abbreviation for* BASE PAIR, BOILING POINT.

brackish a term used to describe water that is intermediate in saltiness, between the sea and fresh water, as occurs in estuaries (*see* ESTUARY).

breccia a sedimentary rock comprising coarse angular clasts (rock fragments). The nature of the rock (i.e. angular fragments reflecting lack of weathering before deposition) implies deposition very close to the source area.

breeder reactor a nuclear reactor capable of manufacturing more nuclear fuel than it consumes while maintaining a CHAIN REACTION.

bremsstrahlung when a charged particle collides with a target and slows rapidly, ELECTROMAGNETIC radiation is emitted as X-rays (*bremsstrahlung*—or "brake radiation"). For example, if an electron collides with a nucleus the KINETIC ENERGY is converted into electromagnetic radiation—X-rays.

bromine (Br) a halogen (group 7A of the PERIODIC TABLE) that exists as a dark red liquid with noxious vapours. Bromine, as with the other halogens, occurs mainly as salts containing the halide ion, and its primary source is SEA WATER. Upon evaporation, NaCl crystals precipitate out of the sea water, leaving a residual fluid (bittern) containing, amongst other compounds, magnesium bromide. Bromine also occurs in marine plants and occasionally in "salt" deposits. It is used in organic chemistry and commercially as a disinfectant, in medicine, and in photography.

bronchus (*plural* **bronchi**) one of two tubular offshoots from the trachea that then split into smaller bronchi and then bronchioles. The bronchi walls are supported by cartilage rings. The bronchioles are very fine tubes that end in alveoli (*singular* alveolus) where carbon dioxide and oxygen are exchanged.

Brönsted-Lowry acid any molecular or ionic substance that will donate PROTONS during a chemical reaction.

Brönsted-Lowry base any molecular or ionic substance that will accept ELECTRONS during a chemical reaction.

Brownian motion a phenomenon first discovered in 1827 by the Scottish botanist Robert Brown (1773- 1858). He observed the random movement of minute particles that occurs in both gases and liquids. Brownian motion is taken as evidence to support the theory that KINETIC ENERGY is an inherent quality of all matter, as it is assumed that

motion that can be seen also occurs in other substances where it cannot be seen.

bubble chamber equipment used to render visible
the paths of ionizing particles. It involves the heating of a liquid (e.g. liquid hydrogen) above its boiling
point and under pressure to prevent boiling. Then,
just before the particles pass through the liquid, the
pressure is released, permitting the formation of
vapour bubbles that "nucleate" on the ions (*see also*
CLOUD CHAMBER).

budding a form of asexual reproduction in which part
of the parent develops a bulge (bud) that becomes
detached to form the offspring. Budding only occurs
in primitive organisms, such as yeast and hydra,
which have a relatively simple structure.

buffer a chemical substance capable of maintaining
the pH of a solution at a fairly constant value. It
does this by removing hydrogen ions (H^+) from the
solution when small amounts of an acid are added,
or releasing hydrogen ions when small amounts of
base are added. Most buffers are ionic compounds,
usually the salt of a weak acid or base.

buffer memory a temporary area of a computer
memory that can be used to store data that has to be
edited or transferred to a disk.

Bunsen burner a gas burner, invented by the German scientist Robert Wilhelm Bunsen (1811–1899)
and used extensively in laboratories, especially in
chemistry. The small upright tube has an adjust-
able air inlet at the base, permitting control of the
flame. The flame produced by burning the hydrocarbon gas/air mix has an inner cone where carbon
monoxide is formed and an outer fringe where it is
burnt.

burette a piece of equipment used for TITRATION in VOLUMETRIC ANALYSIS. It comprises a narrow glass tube, fixed vertically, with a tap at the bottom. The tube is graduated, and small quantities of reagent can be released into a flask, which usually contains a second reagent with an INDICATOR.

butane the fourth member of the ALKANE family (formula C_4H_{10}). It is an extremely useful compound as it is easily changed from a gas to a liquid, thus allowing it to be easily stored and used as a fuel.

byte in computing, a sequence of usually 8 or 16 BITS representing a single character or a unit of memory. The memory capacity of a computer is measured in thousands of bytes (kilobytes, KB) or millions of bytes (megabytes, MB). In the American Standard Code for Information Interchange, (ASCII), 8 bits represent a single character.

C

caffeine a PURINE that occurs in tea leaves, coffee
beans and other plants. It acts as a weak stimulant
to the central nervous system.

calcite ($CaCO_3$) a widespread, common rock-forming mineral, particularly in SEDIMENTARY rocks such
as LIMESTONE, CHALK, and also metamorphosed limestone (MARBLE). It is usually white or colourless and
often is found in crystalline form with good
rhombohedral CLEAVAGE. Well-cleaved rhombs show
double refraction (BIREFRINGENCE). It is used commercially in cement and fertilizer manufacture, and
limestone is used as building stone.

calcium carbonate ($CaCO_3$) an abundant compound
in nature (*see* CALCITE), which is virtually insoluble
in water. Used in the Solvay process for the production of sodium carbonate and ammonia.

calculus a large branch of mathematics dealing with
the manipulation of continuously varying quantities. The many techniques of calculus were developed by the British scientist Isaac Newton (1643-
1727), and independently by the German philosopher Gottfried Wilhelm Leibniz (1646-1716), and
arose from the study of natural phenomena, such as
the changing speed of a falling object. Differential
calculus (*see* DIFFERENTIATION) is concerned with the
rate of change of a dependent variable, i.e. the

maximum and minimum points, gradient, etc, of the graph of a given function. Integral calculus (*see* INTEGRATION) deals with areas and volumes, or methods of summation.

Caledonian throughout geological time there have been periods of mountain building. The Caledonian OROGENY occurred during the late Silurian/early Devonian periods (late Lower Palaeozoic to early Upper Palaeozoic, *see* APPENDIX 5). It affected a vast area including Greenland, Scandinavia, Scotland and Ireland, and was linked with the closure of the IAPETUS OCEAN.

calomel the compound mercury chloride (Hg_2Cl_2), which is used in physical chemistry in a reference electrode. The calomel electrode is a HALF-CELL consisting of a mercury electrode covered with a solution of potassium chloride saturated with calomel.

calorie a unit of quantity of heat defined as that heat required to raise the temperature of one gram of water through 1°C. The calorie has in the main been replaced as a unit by the JOULE (1 calorie = 4.186 joules).

calorimetry the determination of physical constants for substances in physics and chemistry e.g. calorific value, latent heat, specific heat. The technique is based upon the quantitative measurement of heat due to the temperature it produces in a known quantity of water. The **calorimeter** is the instrument used.

Calvin cycle the last stage of PHOTOSYNTHESIS, named after the American biochemist Melvin Calvin (1911–) who discovered that radioactive carbon dioxide ($^{14}CO_2$) became incorporated into the carbohydrate subsequently found in the cells of plants.

All the various chemical reactions involved in the Calvin cycle occur in the stroma of plant cells. These reactions are termed the DARK REACTIONS of photosynthesis, since the formation of glucose can occur without light, although it does need the products generated by the light reactions, e.g. ATP. The fixation of carbon dioxide and its conversion into carbohydrate is an energy-requiring process, as it takes six turns of the Calvin cycle to produce one molecule of glucose. The required energy is generated by the HYDROLYSIS of ATP.

Campbell-Stokes recorder *see* **sunshine recorder**.

cancer a disease characterized by an uncontrolled growth rate of cells, leading to the formation of tumours. If the tumour remains localized, it is termed benign and is usually harmless to the host. However, malignant tumours do not remain localized but, instead, spread throughout the body and set up a secondary growth area by a process called METASTASIS. This usually causes death in the host as the malignant tumour disrupts the essential everyday processes of the cells in the affected tissues. There is not a single cause of cancer—it is triggered by a combination of factors, including exposure to carcinogens (such as tobacco smoke), radiation, ultraviolet light, certain viruses, and the possible presence of potential cancer genes (oncogenes). Treatment of different cancers involves surgery, radiotherapy, chemotherapy and, just recently in 1987, the "magic bullet" approach, where cytotoxic drugs are labelled with one specific ANTIBODY, which will only recognize a protein present on the surface of cancer cells.

candela the SI unit of luminous intensity.

capacitance the ratio of electric charge stored within a system in relation to its electric potential. A capacitor stores electric charge (Q) according to its capacitance (C) when its potential (V) has a value of one volt. Thus a system has a capacitance of one FARAD (the unit of capacitance) when a charge (Q) of one COULOMB changes the potential by one volt (V), i.e. $C = Q/V$.

capacitor a device used to store electric charge, comprising thin conductors separated by insulating, DIELECTRIC material, e.g. MICA, glass, or simply a vacuum. The maximum p.d. (POTENTIAL DIFFERENCE) that can be applied is dependent upon the dielectric used and when it breaks down.

capillary action a phenomenon related to SURFACE TENSION in liquids and due to inter-molecular attraction at the boundary of a liquid, which results in liquid rising or falling in a narrow tube.

capillary pressure (*see* CAPILLARY ACTION) for a capillary tube with radius r, and a liquid of surface tension γ with an angle of contact θ, the capillary pressure is defined as $P = 2\gamma \cos \theta/r$.

cap rock an impervious rock overlying a source rock containing hydrocarbons. The cap rock contains the oil or gas and prevents upward migration. Typical cap rocks are limestone, shale, evaporite, and clay-rich sandstone.

carat (1) a standard weight used for measuring stones, and now equal to 200 mg. (2) a measure of the fineness of gold where pure gold is 24 carats and 22 carat gold contains 2 parts alloy.

carbides compounds of metals with carbon which in many cases produce hard, refractory materials with metallic conductivity, e.g. TUNGSTEN carbide. Carbide tools are made of tungsten or tantalum carbide, or mixtures with nickel and cobalt and are ideal as cutting tools for hard materials at high temperatures.

carbohydrates a large group of compounds containing carbon, hydrogen and oxygen, with the general formula $C_x(H_2O)_y$. The group includes the sugars, starch, and cellulose, and form the mono-, oligo- and polysaccharides. Carbohydrates play a vital role in the metabolism of all living organisms.

carbolic acid *see* **phenol**.

carbon a non-metallic element that exhibits ALLOTROPY, occurring as DIAMOND and graphite and amorphous carbon black. It is used extensively as motor brushes, in steelmaking, and in the manufacture of cathode-ray tubes. Carbon is unique in the enormous number of compounds it can form, and is the basis of all organic chemistry and thus essential to all living organisms (*see* CARBON CYCLE).

carbonaceous rocks sedimentary rocks that contain carbon (derived from plant matter) in some form as the main constituent. Included are lignite, brown coal, and the various forms of true coal.

carbonate(s) compounds containing the CO_3 group. In geology, a group of minerals occurring primarily in LIMESTONES and DOLOMITES. CALCITE ($CaCO_3$) is the most important, aragonite (also $CaCO_3$) is a less stable form and DOLOMITE contains magnesium.

"Carbonates" is also used as a general collective term for sedimentary rocks with $\geq$ 95 per cent calcite or dolomite and usually refers to limestone.

carbon compounds based upon carbon (C), element 6 in the PERIODIC TABLE, these include all organic compounds and are the basis of all living matter.

carbon cycle the circulation of carbon compounds in the natural world by various metabolic processes of many organisms. The main steps of the carbon cycle are:

(1) Carbon dioxide present in air and water is taken up during photosynthesis in plants and some bacteria.

(2) The carbon accumulated in plants is later released during the decomposition of the dead plant, or of bacteria or animals that have consumed any of the plant.

(3) Carbon will also be released by the burning of fossilized plants in the form of fuels—coal, oil and gas—and during the respiration of all organisms. The concentration of carbon dioxide in the atmosphere is increasing as huge areas of tropical forests are destroyed, while the consumption of fossil fuels is rising, i.e. less PHOTOSYNTHESIS to absorb the increasing CO_2 level. This may be a factor involved in the small temperature rises throughout the world, known as the GREENHOUSE EFFECT. When there are high levels of carbon dioxide in the atmosphere, heat radiation from the Sun tends to be reflected back to Earth rather than lost to space.

carbon dioxide (CO_2) a colourless gas occurring in the ATMOSPHERE due to OXIDATION of carbon and CARBON COMPOUNDS. It is the source of carbon for plants and plays a vital role in METABOLISM. It solidifies at -78.5°C and is much used as a refrigerant. It is also used in carbonated drinks and, since it does not support combustion, in fire extinguishers.

carbon fibres a very strong (for its weight) fibre that is made by the heat-treatment of organic fibres, which results in all side chains being removed to leave the carbon chain. The chains are orientated along the fibre, producing a final diameter of 7 or 8μm. Carbon fibres are then used as reinforcement in composites with resins, ceramics, etc, resulting in a strong material that will withstand high temperatures.

carbonic acid (H_2CO_3) a short-lived weak acid formed by the dissolution of CARBON DIOXIDE in water. When obtained in a chemical reaction, however, it breaks up into its constituent molecules. Carbonic acid gas is the carbon dioxide in liquids that is dissolved under pressure and then released in effervescing bubbles when the restraint is removed.

carbon monoxide (CO) a colourless gas formed during the incomplete combustion of coke and similar fuels. It also occurs in the exhaust fumes of motor engines. Carbon monoxide is poisonous when breathed in because it combines with the HAEMO-GLOBIN in the BLOOD to form a stable compound. This reduces the oxygen-carrying capacity of the blood. It is a valuable industrial reagent, due to its reducing properties.

carbon tetrachloride *or* **tetrachloromethane** (CCl_4) until recently, extensively used as a solvent for fats and oils, but its use in dry cleaning (and also fire extinguishers) has fallen considerably due to its toxicity. It is a colourless liquid, formed by the chlorination of METHANE.

carboxyl the acid group -COOH, characteristic of the carboxylic acids where an oxygen is double-bonded and the hydroxyl (OH) is singly bonded to the carbon.

carboxylic acid an organic acid containing one or more CARBOXYL groups, e.g. propanoic acid, CH_3CH_2COOH.

carcinogen a substance that may produce CANCER (carcinoma).

cardiovascular system the organization of the heart, the arteries and veins within the human body, in which the heart and the blood vessels form a virtually closed system. The minor branches of the blood vessels supply blood to every part of the body, including the bones. The only bloodless parts of the body are dead structures such as the nails and hair. However, the nail-bed and the hair roots do require a blood supply. The only other organ without its own blood supply is the cornea (the clear window of the eye). The cornea is supplied with oxygen and nutrients by means of the tears. Other than these exceptions, every part of the body requires a constant blood supply in order to receive essential nutrients such as GLUCOSE, AMINO ACIDS, etc. Any interruption in the blood supply of the cardiovascular system causes death of that tissue.

carotenoids plant pigments, coloured orange, red and yellow, that chemically resemble TERPENES and are based upon the tetraterpene unit ($C_{40}H_{64}$). The group includes the carotenes, which are simple hydrocarbons, and the xanthophylls (oxygenated derivatives of carotenes). The characteristic colours of, for example, carrots and ripe tomatoes are due to carotenoids. Although they are not essential to PHOTOSYNTHESIS, they absorb photons and the energy is transferred to the CHLOROPHYLL. In addition, these accessory pigments protect cells in some bacteria from photo-oxidation.

carnivore any animal that eats the flesh of other animals. The term can also refer to the mammalian group *Carnivora*, which includes bears, cats and dogs. Carnivorous plants are ones that trap and digest insects (using special ENZYMES) in order to obtain their nitrogen requirements.

carpel the female reproductive organ of a flower, comprising stigma, style and ovary. The ovary holds one or more ovules, which form seeds after fertilization.

Cartesian co-ordinates a method of representing the position of a point in space, invented by the French mathematician and philosopher, René Descartes (1596-1650). For instance, the point with Cartesian coordinates (5,2) will be found by moving 5 units along the horizontal (X-AXIS) and 2 units up the vertical axis (Y-AXIS).

cartography the evaluation and compilation of all forms of data for, and the design and draughting of, a new or revised map. Cartography also includes the study of maps, methods of presentation, and map use.

casein any of a group of proteins containing phosphate, found in milk and cheese. It is precipitated as curd by the action of rennin (an ENZYME) and calcium.

Castner-Kellner process the production of sodium hydroxide, chlorine and hydrogen by the ELECTROLYSIS of brine. Sodium is produced at and dissolves in a mercury cathode. The resulting AMALGAM is removed. and through the action of water, sodium hydroxide and mercury are formed.

catabolism the degradational processes of METABOLISM. For example, the breakdown of glucose during GLYCOLYSIS is a catabolic process.

cataclasis deformation of rocks by shearing and the breaking of mineral grains into smaller pieces (granulation) with little or no chemical change. The resulting rocks, **cataclasites**, are found in areas of tectonic activity and shear zones (*see also* MYLONITE).

catalyst a substance that increases the rate of a chemical reaction but can be recovered unchanged at the end of the reaction. Metal catalysts such as iron and platinum are used throughout industry to lower the ACTIVATION ENERGY of a specific process. All biological processes in animals and plants involve natural catalysts called ENZYMES.

cathode the negatively charged electrode of an electrochemical cell to which CATIONS travel and gain electrons (i.e. where REDUCTION occurs).

cathode rays a stream of electrons emitted from the negative electrode (*see above*) when electricity is passed through a VACUUM TUBE. In a piece of equipment known as the cathode-ray oscilloscope, the electrons emitted by the cathode are attracted to and hence accelerated by an anode (positive electrode). The electrons are accelerated farther down the vacuum tube between two sets of parallel plates and onto a fluorescent detecting screen. This equipment is used in televisions, in computers and in laboratories to measure the frequencies of waveforms.

cathodoluminescence luminescence induced by the irradiation of a material with electrons. It is used in the study of minerals, in particular to differentiate between cements or stages of mineral growth in SEDIMENTARY ROCKS.

cation a positively charged ion formed by an atom or group of atoms that has lost one or more electrons.

celestial body any of the stars or planets within that part of space studied in astronomy.

celestial equator the circle created by the intersection of the plane of the Earth's equator and the CELESTIAL SPHERE.

celestial mechanics a part of astronomy including the movement of celestial bodies (or systems) influenced by gravity. Isaac Newton (1643–1727) developed this subject; in the main, it deals with the movements of planets and satellites within the solar system.

celestial poles the points at which northerly and southerly extensions of the Earth's axis meet the CELESTIAL SPHERE.

celestial sphere the imaginary sphere that has the observer at its centre and all stars, irrespective of their true distance, are placed on the inner surface of the sphere, referenced by co-ordinates.

cell the basic unit of all living organisms. In most cases, cells are microscopic and may form unicellular organisms as in bacteria (*see* BACTERIUM), or merge to create tissues or colonies. The two types of cell are PROCARYOTE and EUCARYOTE, the former being the more primitive.

cellulose a polysaccharide occurring widely as cell walls in plants. It is found in wood, cotton and other fibrous materials and comprises chains of GLUCOSE units. It is used in the manufacture of paper, plastics and explosives.

Celsius scale *or* **centigrade scale** (C) a temperature scale with a freezing point of 0° and a boiling point of 100°, devised by the Swedish astronomer Anders Celsius (1701-44).

central processing unit *see* **CPU**.

centre of gravity gravity acts upon all parts of a body, and the sum effect of these forces acts through a single point, called the centre of gravity. In a uniform gravitational field, this coincides with the centre of mass, i.e. the point where the mass of a body is considered to be.

centre of symmetry a concept of symmetry used in crystallography, the centre of symmetry is a central point around which all crystal features occur in equidistant and equal pairs, i.e. like faces, edges, etc, are arranged in pairs in corresponding positions on opposite sides of the central point. The cube has an obvious centre of symmetry.

centrifugation a technique in which a high-speed rotating machine, a centrifuge, generates centrifugal forces to separate the various components of a liquid. Different components suspended in the liquid will separate at different centrifugal speeds depending upon their size and mass.

centripetal force the product of a body's mass and the centripetal acceleration. The centripetal acceleration is the acceleration of a body moving along a curved path, directed towards the centre of that path's curvature. The centripetal force is equal and opposite to the centrifugal force.

centromere the constricted region of a CHROMOSOME, to which a pair of sister CHROMATIDS are attached. All nuclear chromosomes of EUCARYOTIC cells have a centromere, the exact position of which varies but which is genetically inheritable. The centromere can therefore be used in the identification of chromosomes, which, to be observed under the microscope, must be in cells which are undergoing either MEIOSIS or MITOSIS. The centromere is also an important factor in nuclear division as the absence of one will lead to failure of segregation; that is, one daughter cell will contain both sister chromosomes instead of one chromosome per each separate daughter cell. Thus only one cell is inheriting the genes that code for vital proteins.

CFC (*abbreviation for* chlorofluorocarbon) a chemical widely used in manufacturing processes, which reacts with and destroys OZONE, causing depletion of the OZONE LAYER.

chain reaction a multi-step reaction in which the products of each step are reactants in the subsequent step. The mechanism for a chain reaction involves a slow initiation reaction, followed by chain propagation until the reaction becomes inhibited and terminates.

chain rule a method used in CALCULUS to differentiate a FUNCTION of a function. By substituting another differentiable function into the original function, a COMPOSITE function is formed and differentiation can proceed using the following formula:

$$\frac{dy}{dx} = \frac{dy}{du} \times \frac{du}{dx}$$

chalcedony a type of QUARTZ comprising minute crystals too small to be seen even with a microscope. It occurs in some sedimentary rocks and filling vesicles in lavas. Chert is a nodular form and in chalk occurs as FLINT.

chalk a white, fine-grained, porous limestone formed from CALCIUM CARBONATE and calcareous skeletal remains of micro-organisms. Chalk deposited in the Upper Cretaceous (*see* APPENDIX 5) covers much of northwest Europe.

Charles' law this states that at constant pressure, the volume of a gas varies directly with the temperature, i.e. if the temperature is doubled, then the gas volume will also double (*see also* GAS LAWS).

chelation a reaction between a metal ION and an ORGANIC molecule which produces a closed ring thus tying up the unwanted metal ion. Chelation occurs naturally in soils, removing metal ions in solution which may be potentially toxic to plants. This principle can be applied to domestic products, e.g. chelating agents are often added to shampoos to soften water by "locking up" calcium, iron and magnesium ions (*see* HARD WATER.)

chemical bond *see* BOND.

chemical equation the representation of a chemical reaction using SYMBOLS for atoms and molecules.

chemical equivalent *see* **equivalent weight**.

chemical oxygen demand (COD) an indicator of water (or effluent) quality. Oxygen demand is measured chemically (*compare* BIOLOGICAL OXYGEN DEMAND) using potassium dichromate as the oxidizing agent. The OXIDATION takes two hours, providing a much quicker assessment than the BOD.

chemisorption *see* **adsorption**.

chemistry the study of the composition of substances, their effects upon one another and the changes which they undergo. The three main branches of chemistry are ORGANIC, INORGANIC, and PHYSICAL chemistry.

chemotaxis the movement of an organism towards or away from a specific chemical. For example, some bacteria in a solution have been shown to move (using their FLAGELLA) towards an area of high glucose concentration.

chemotherapy the use of toxic chemical substances to treat diseases where the chemicals are directed against the invading organisms or the abnormal tissue.

chimaera an individual with tissues of two or more different GENOTYPES. The cause may be mutation in a developing embryo. In plants it may be due to mutation or grafting.

china clay a clay, composed primarily of KAOLINITE, which is formed due to hydrothermal alteration of granite. It is extracted using high pressure water jets and is used in many industries including ceramics, paper and pharmaceuticals.

chip a tiny piece of semiconducting material, such as SILICON, printed with a microcircuit and used as part of an integrated circuit.

chirality in STEREOISOMERISM, the idea of left- and right-handedness applied to chemical molecules (*see also* OPTICAL ACTIVITY). The structure of a molecule is **chiral** if it cannot be superimposed on its own mirror image.

In mathematics, the application of right- and left-handed to co-ordinate axes.

chi-squared test a statistical test used to determine how well data obtained from an experiment (the observed data) fits with the data expected to occur by chance. The chi-squared test is a simple method of checking that the experimental results are significant and have not just arisen from chance events.

chitin a HYDROCARBON, related to CELLULOSE but containing nitrogen, which forms the skeleton in many invertebrates, e.g. the shells of insects.

chlorine the second of the halogens (group 7A of the PERIODIC TABLE) that exists as a yellow/green gas. It is a choking irritant, which has a harmful effect if inhaled. It occurs widely as the chloride, the commonest being salt, NaCl. It is manufactured primarily by the ELECTROLYSIS of brine and is itself used extensively. Chlorine is a powerful oxidizing agent and is used in the production of bleaches, disinfectant, and hydrochloric acid. It is also used in the production of organic chemicals, e.g. CARBON TETRACHLORIDE, PVC, and a variety of plastics and solvents.

chlorofluorocarbon *see* **CFC, ozone layer**

chlorophyll the green pigments of plants present in the cell ORGANELLES called CHLOROPLASTS. Chlorophyll is an essential factor in PHOTOSYNTHESIS as it traps energy from sunlight and uses it to split water molecules into hydrogen and oxygen.

chloroplast an ORGANELLE which is found within the cells of green plants and ALGAE and which is the site of PHOTOSYNTHESIS. The chloroplast consists of closed membrane discs called thylakoid vesicles (often stacked together to form a granum), which are surrounded by a watery matrix termed the stroma. The stroma contains enzymes necessary for the CALVIN CYCLE, and the thylakoid vesicles contain chlorophyll, an essential pigment of photosynthesis as it absorbs light energy.

cholesterol an insoluble molecule that is abundant in the plasma membrane of animal cells. In mammals, cholesterol is synthesized from saturated fatty acids in the liver and is transported by carrier molecules in the blood, called low-density LIPOPROTEIN (LDL). Cholesterol enters cells via the LDL-receptor on the plasma membrane of the cell. This mechanism of uptake helps regulate the levels of cholesterol in the blood. If a person's diet is high in cholesterol, the number of LDL-receptors on the cell membrane will decrease, which will result in a decrease in the cellular uptake of cholesterol leading to a corresponding increase in the levels of cholesterol in the blood. This excess cholesterol is deposited in the artery walls and ultimately leads to ARTERIOSCLEROSIS. Some unfortunate individuals are particularly prone to arteriosclerosis as they carry a gene that codes for a defective LDL-receptor, making the cellular uptake of cholesterol an impossible task. This inherited disorder is called familial *hypercholesterolemia*, which translates as "too much cholesterol in the blood." Cholesterol is also the main component of steroid hormones and bile salts.

choline a base ($CH_2OHCH_2N(CH_3)_3OH$) that occurs in certain types of PHOSPHOLIPIDS. It is found in the brain, egg yolk, and bile in this form. In the form of ACETYLCHOLINE, it is an important substance in the transmission of nerve impulses (*see also* SYNAPSE).

chromatid one of a pair of side-by-side replica CHROMOSOMES joined by the CENTROMERE and produced during the replication of DNA within a cell.

chromatography a method for isolating the constituents of a solution by exploiting the different bonding properties of molecular substances. One complex method is affinity chromatography, which is used to separate proteins. A specific ANTIBODY is coupled to small plastic beads to form a column through which the protein solution is passed. The only protein that will bind to the column is the one recognized by the antibody; other proteins will pass through the column unimpeded.

chromosome a highly structured complex of DNA and PROTEINS found within the nucleus of EUCARYOTES. A nuclear chromosome is involved in the following three functions:

(1) Replication—this ensures the correct duplication of the DNA, or cells would become unviable through genetic mutation.

(2) Segregation—this ensures that the newly replicated chromosomes will separate and that each will become part of a daughter cell. It is a very important function dependent largely on the CENTROMERE.

(3) Expression—this ensures that the genes present on the chromosome are correctly transcribed to preserve the genetic information coding for specific proteins.

PROCARYOTIC cells contain a single, circular, chromosomal DNA molecule, whereas evidence indicates that EUCARYOTIC cells contain a single, linear molecule of double-helical DNA within their nucleus. However, the eucaryotic ORGANELLES, MITOCHONDRIA and CHLOROPLASTS contain a single circular molecule of DNA, which perhaps is an indication of their origin, i.e. as procaryotic cells that have evolved into eucaryotic organelles. A species will have a characteristic number of base pairs of DNA all packed into a characteristic number of chromosomes. For instance, yeast cells contain approximately 14 x 106 base pairs (bp) packed into 16 chromosomes, and the human nucleus contains approximately 3000 x 106 bp packed into 46 chromosomes. Chromosomes are visible under the microscope only during MEIOSIS or MITOSIS, when they condense to form short thick structures before separating.

chromosphere the gaseous layer surrounding the SUN, which is visible during a total ECLIPSE. It surrounds the PHOTOSPHERE and is several thousand miles thick.

cilia (*singular* **cilium**) fine thread-like structures on cell surfaces, which beat to create currents of liquid over the cell surface or to move the cell.

circadian (rhythm) regular changes occurring in the activity, behaviour or physiology of animals or plants approximately every 24 hours. The sequence may continue for days even in the absence of normal, repetitive events, e.g. daylight and darkness, with which it is linked.

circuit a path that, when complete, allows electric charge to flow through it. There are only two possible types of circuit configurations—PARALLEL CIRCUITS and SERIES CIRCUITS. In series circuits, the same amount of current flows through all the components, but there is a different drop in POTENTIAL DIFFERENCE through each component. In parallel circuits, a different amount of current flows through each component, but the same drop in potential difference occurs across each component of the circuit.

cirrocumulus a type of high-level cloud formed in stable air as sheets or layers with ripples and waves.

cirrostratus a type of cloud composed of whitish or near-transparent veils with a smooth or fibrous form. Often this cloud type covers the sky, producing a rainbow or white ring around the SUN or MOON due to refraction by ice crystals.

cirrus a high-level cloud forming narrow bands or filaments.

citric acid $C_3H_5O(COOH)_3$ a tricarboxylic acid that occurs in many fruits, especially lemons. It is now produced commercially by fermentation of sugar by moulds of the *Aspergillus* genus and is itself used as a food flavouring and in the production of effervescent salts and drinks.

citric acid cycle a complex set of biochemical reactions controlled by ENZYMES. The reactions occur within living cells, producing energy, and the cycle is instrumental in the final stages of the OXIDATION of CARBOHYDRATES and fats and is also involved in the synthesis of some AMINO ACIDS.

clathrate a compound in which molecules of one type are enclosed in the structure of another different molecule. A property that is used in the separation of gases (*see also* ZEOLITES).

clay a fine-grained sediment composed of clay minerals that are hydrous aluminium silicates. Clay is used extensively in numerous industries, including the manufacture of ceramics and bricks; rubber, plastic and paint production; as fillers in paper manufacture, and in drilling muds.

cleavage in minerals, the tendency to split along definite planes. These cleavage planes reflect the internal CRYSTAL structure, and their directions are parallel to a face or faces of a form in which the mineral may crystallize. This plane of splitting exhibits the least cohesion, due to the atomic structure of the crystal, and while minerals may have several cleavages, one is usually more readily shown. In rocks, cleavage includes structures of diverse origin, the common factor being the fissility that allows the rock to be split. The main types are slaty cleavage, shown best in fine-grained metamorphosed mudstone (i.e. slates) where alignment of flaky minerals (MICA, clay minerals) creates new parallel planes; fracture cleavage, which is a set of closely spaced fractures typical of deformed, relatively strong rocks such as SANDSTONES; crenulation cleavage, which, as the name suggests, is due to very small folds (crenulations) of thin layers in a rock. The alignment of closely spaced folds produces a marked cleavage or FOLIATION (*see also* SCHIST).

climatic zones regions of the Earth possessing a distinct pattern of climate, each region approximating to belts of latitude. There are essentially eight zones. Working from the two polar caps of snow, there is a zone of boreal (northern) climate in the northern hemisphere with a range of temperatures, two belts of humid temperate conditions, two subtropical arid or semi arid zones, and finally the humid tropical climate near the equator.

clone an organism, micro-organism or cell derived from one individual by asexual reproduction. All such individuals have the same GENOTYPE. Plants propagated by cuttings are clones.

closed-chain in organic chemistry when carbon atoms are bonded together to form a ring, as typified by BENZENE. Closed-chain compounds are thus called ring or cyclic compounds (*see also* HETEROCYCLIC COMPOUNDS).

cloud water droplets or ice formed by condensation of moisture in rising warm air. Rising on convection currents, the warm moist air reaches the higher, cooler conditions and condenses. Clouds are classified on their form into three major groups—cumulus (heap), stratus (sheet) and cirrus (fibrous)—which are further subdivided into genera and species on features such as transparency, type of growth, and arrangement.

cloud chamber *or* **Wilson cloud chamber** an instrument that renders visible a stream of charged particles moving through a gas. A chamber contains a saturated vapour that is expanded (and therefore cooled) and that, on passing through the beam of particles, creates ions upon which the vapour condenses. The trace of the particles is therefore revealed by the liquid droplets (*see also* BUBBLE CHAMBER).

coal *see* FOSSIL FUELS.

coal-tar one by-product from the decomposition of coal by heat in the absence of air (i.e. carbonization, a process for the production of coke, town gas, and smokeless fuel). Depending upon the temperature of carbonization, coal-tar contains a cocktail of chemicals including BENZENE, toluene, PHENOLS and pyridine.

COBOL (*acronym for* Common Business Orientated Language) a programming language geared towards a computer, which is mostly used in commerce.

COD *abbreviation for* CHEMICAL OXYGEN DEMAND.

coefficient the constant numerical part of a term in an algebraic equation. For example, in the equation $5x - 6xy + y = 0$, the coefficient of x is 5, that of xy is -6, and that of y is 1. In some mathematical expressions, unknown coefficients can be represented by letters and their values calculated using a particular formula.

coelostat an instrument that is used in conjunction with an astronomical telescope to reflect the same region of the sky into the telescope.

coherence the state when a beam of light or other electromagnetic radiation has waves that are in phase at the same part of the cycle.

cohesion the attraction between molecules in a liquid that enables thin films and drops to form.

cold front the line defining the front of a cold air mass that will cut under warm air ahead of it. The weather changes that may occur include a fall in the temperature and veering often squally wind.

collagen an important fibrous PROTEIN that forms almost one-third of total body protein in mammals. It is found in tendons, bone, skin and cartilage, and comprises mainly glycine, proline (AMINO ACIDS) and hydroxy-proline in helical structures.

colligative property an aspect of a solution that depends solely on the concentration and not the nature of the dissolved particles. Colligative properties are important when determining, say, osmotic pressure (*see* OSMOSIS), vapour pressure, or the freezing point of a solution.

collimator a device used to produce a parallel (or nearly so) beam of particles or radiation, comprising a slit that is at the focus of a lens. The radiation or light entering the slit leaves the collimator as a parallel beam.

collision in physics, an interaction between particles where MOMENTUM is maintained. The collision is termed elastic if the KINETIC ENERGY of the particles is also conserved, and inelastic if not. In nuclear physics, collision tends to mean a proximity that allows an interaction due to the forces associated with these particles. An "actual" collision would result in capture of a particle.

colloid a substance forming particles in a solution varying in size from a true solution to a coarse suspension. The particles, measuring 10^{-4} to 10^{-6} mm are charged and can be subjected to ELECTROPHORESIS.

colony a group of individuals of the same species living together and, to some extent, interdependent. In some cases the individuals are actually joined, as in corals, and function as one. In others, insects for example, they are separate but exhibit a high degree of organization, possibly with specialization of functions. Also, a colony may be a BACTERIUM or yeast growing from a parent cell on a food source.

colorimeter an apparatus used to measure the hue, brightness and purity of a colour. In colorimetric analysis, the colour of a solution is analysed to enable comparison with a known standard, thus permitting quantitative analysis of the solution.

combination the random selection of a group of objects from a given set to form a subset. The number of combinations for selecting a specific size of subset can be calculated using:

$$nC = \frac{(n)}{(r)} = \frac{n!}{r!\,(n-r)!}$$

where r = size of subset to be selected;
n = size of original set;
and ! = factorial $(3! = 3 \times 2 \times 1 = 6)$

For example, if a coach wanted to select 15 players out of a total of 20 to form an American football team, there would be

$$\frac{20!}{15! \times 5!} = 15504 \text{ (ways of selecting the team)}$$

comet a small body moving in eccentric orbit around the Sun and composed of dust, gas and frozen ices (of H_2O, CO_2, CO, HCHO). A comet is formed of a head and tail. The former comprises the nucleus and the coma, which is a layer of gas. The nucleus is irregularly shaped and a few kilometres in diameter. The tail is formed by the emission of gas and dust when within a few ASTRONOMICAL UNITS of the Sun and points away from the Sun due to repulsion by the SOLAR WIND and radiation.

commensalism where individuals from different species live together such that one benefits from the association while the other is not affected.

commutative a term used to describe any mathematical operation that will give the same result independent of the order in which the elements of the operation are performed. Thus, multiplication is commutative, e.g. $2 \times 3 \times 6 = 6 \times 2 \times 3$, but subtraction is not, e.g. $6 - 3 \neq 3 - 6$.

commutator the part in a motor or ARMATURE that makes contact with the carbon brushes to carry current.

common difference the value given to the difference between successive terms in an ARITHMETIC SERIES.

common logarithm *see* **logarithm**.

complementary angles two angles totalling 90°. With such angles one is the complement of the other.

complex number a number represented by the symbol z, which has the following form:

$$z = a + ib$$

(where a and b are real numbers and i is $\sqrt{-1}$).

component when resolving forces, two forces acting in different directions are substituted for the single, original force and the latter is equal to the resultant of the two components. Resolving into components can apply to acceleration, velocity, etc, and solution is achieved using TRIGONOMETRIC FUNCTIONS.

composite a term used to describe a function, number or POLYNOMIAL that has factors that differ from the given function, number or polynomial. For example,

73 x 5 = 365 is composite and y = x (x + 8)3 is a composite function. PRIME NUMBERS are not composite as they are divisible only by themselves or unity.

composition of functions the mathematical operation in which a single function is formed from two given functions. The function is formed by inserting the second function into the first equation. In mathematical terminology, the composition of function "f" with function "g" is written as $f \circ g$. For example, the two functions, $f(x) = x + 2$, and $g(x) = x^2$ gives the following:

$$f \circ g = f[g(x)] \text{ or } g \circ f = g[f(x)]$$
$$= g + 2 \qquad\qquad = f^2$$
$$= x^2 + 2 \qquad\qquad = (x + 2)^2$$

concentration the quantity of substance (SOLUTE) dissolved in a fixed amount of SOLVENT to form a SOLUTION. Concentration is measured in moles per litre (mol l^{-1}).

concentric a mathematical term used to describe geometrical figures that share a common centre.

concretion a nodular concentration formed from solution and found in SEDIMENTARY ROCKS. Commonly siliceous or calcareous, it is formed by localized deposition. The term also applies, in the study of soils, to concentration of, for example, iron oxide.

condensation the process by which a substance changes from the gaseous state to the liquid state and, in so doing, loses KINETIC ENERGY.

The process, in meteorology, of forming a liquid from its vapour. If moist air is cooled below its DEW-POINT and nuclei or surfaces are available then water vapour will condense. The nuclei may be particles of dust or ions.

In chemistry, the reaction of one molecule with

another, and the elimination of a simple molecule such as water or an alcohol.

condensation trails cloud created by an aircraft. It is formed by the CONDENSATION caused by pressure reduction above the wings, or by condensation of water vapour in exhaust gases.

condenser in chemistry, a piece of apparatus used to condense vapours, usually tubular and cooled by a water filled jacket.

In physics, a large lens or mirror that collects and directs the light in a projection system onto the transparency or other object to be focused by the projection lens.

conductance the reciprocal of RESISTANCE in a circuit. The unit was the mho (reciprocal ohm), but the SI unit is the siemen (S).

conduction a method of heat transfer through a solid, or the flow of electrical charge through a substance. Transfer of heat energy by conduction is always from a high temperature to a lower temperature region. A poor **conductor** of heat and electricity is called an INSULATOR, e.g. plastic, cork. All metals and carbon (GRAPHITE form) are good conductors of electric charge as they contain ELECTRONS, which are free to carry and thus transfer energy (*compare* ELECTROSTATICS).

conductivity the reciprocal of RESISTIVITY with units siemens m^{-1}.

conductor *see* **conduction**.

conglomerate a coarse-grained rock containing clasts (rock fragments) that are rounded or sub-rounded and larger than 2 mm (0.079 in).

congruent in mathematics, a term used to describe the relationship between figures having an identical shape and size, but possibly a different orientation in space. It can also refer to a pair of numbers relative to a third number, which is indeed the difference or a multiple of the difference, between the original pair.

conservation of energy, law of *see* **energy, thermodynamics**.

constant in equations or algebraic expressions, any quantity that remains the same.

constellation a group of stars represented pictorially and allocated a name that, although commonly used, has no scientific import.

constructive boundary *see* **divergent junction**.

contact process the inexpensive manufacture of the important industrial chemical SULPHURIC ACID (H_2SO_4). Sulphur or sulphur ores are oxidized to obtain sulphur dioxide (SO_2), which is further catalytically oxidized to sulphur trioxide (SO_3). The extremely reactive SO_3 gas is then dissolved in concentrated sulphuric acid (H_2SO_4), and the resulting pyrosulphuric acid ($H_2S_2O_7$) is diluted with a specific volume of water to give the desired concentration of sulphuric acid (H_2SO_4).

continental crust the Earth's crust that lies beneath the continents and CONTINENTAL SHELVES. It is 30–40 km (19–25 miles) thick mostly, but increases to approximately 70 km (44 miles) beneath areas of mountain building. The crust has two layers of differing compositions and slightly different densities, and its base is marked by the MOHOROVICIC DISCONTINUITY, where both composition and density change significantly.

continental drift a geological concept, formulated by the German geophysicist Alfred Wegener (1880–

1930), that 200 million years ago the Earth consisted of a large single continent, called PANGAEA, which broke apart to form the present continents. An explanation of how such huge land masses move is provided by studying the vast plates that make up the outer layer of the Earth, called the crust. The crustal plates are believed to float on a partially molten region of the Earth between the crust and the Earth's core, the lower mantle (hence the modern term plate tectonics). *See also* SEA-FLOOR SPREADING and MID-OCEANIC RIDGE.

continental shelf the surface between the shoreline and the top of the continental slope, where the gradient steepens at a depth of approximately 150 metres (*c.*490 ft). The average width of the shelf is 70 metres (*c.*230 ft).

contour a line connecting points on the surface that are the same height above datum (reference surface).

convection a method of heat transfer through liquids and gases. It will only occur if the lower temperature area is above the high temperature area of a liquid or gas. Convection currents can be seen easily when a dye is placed at the bottom of a container of water and heat is applied to that region, causing the dye and water molecules to rise and disperse throughout the container. An everyday example of convection that can be readily detected by hand is the hot air that rises from a warm radiator.

In geology, convection is postulated as being the driving force, within the MANTLE, for PLATE TECTONIC motion. Heat released by nuclear disintegration creates convective cells throughout the mantle and depending upon the site, creates a number of mechanisms for the different plate movements.

convergence zone, boundary, *or* **margin** the boundary where two tectonic plates meet, resulting in one of two effects. One creates a SUBDUCTION ZONE that on the Earth's surface follows deep ocean trenches. In this case, the OCEANIC CRUST is pushed down beneath CONTINENTAL CRUST, and since this returns old LITHOSPHERE material to the mantle, it is termed a destructive plate margin. The other case is that of a continental collision, where all the oceanic crust has been subducted and the two continents eventually lock together.

co-ordination compound a compound with atoms or LIGANDS arranged around a central atom, where the ligand donates electrons in forming a co-ordinate (COVALENT) bond with the central atom. The central atom is often a transition metal ion (*see* TRANSITION ELEMENT), and the ligands are negatively charged.

copolymer a polymer compound formed by polymerizing two or more MONOMERS. A number are important industrially and result from the polymerization of vinyl chloride with vinyl acetate, and styrene with butadiene.

coral reef a hard bank built up by the carbonate skeletons of colonial corals (and algae). There are several forms of reef, including BARRIER REEFS, fringing reefs (which are attached to the coast) and atolls (where a reef encloses a lagoon). Certain conditions are essential for the growth of reefs, including temperature, a maximum water depth of 10 metres (33 ft), clear water with no land-derived sediment, and normal marine salinity.

Coriolis force when dealing with moving objects in relation to rotating systems (e.g. a particle moving away from an observer on the Earth), although the particle moves in a straight line, it will appear to the observer to move in a curved path. The Coriolis force is a theoretical force that is used to simplify such calculations, e.g. in calculating movement of air on the Earth's surface.

corona in meteorology, the coloured rings (observed through thin cloud) around the Sun or Moon, caused by diffraction of light by water droplets (*see also* AUREOLE). Also in physics, when air breaks down at the surface of a conductor due to electric stress, the point before discharge.

coronary artery one of the two main vessels that carries oxygenated blood to the heart muscle. In general terms, the left coronary ARTERY divides into two branches and supplies most of the blood required by the left VENTRICLE, while the right coronary artery supplies blood to the right ventricle. The two coronary arteries originate from the AORTA, but run independently on to the surface of the heart.

corrosion the gradual wearing away of solids (particularly metals) by chemical attack.

corundum a hard mineral oxide (Al_2O_3) of aluminium. The two main varieties are sapphire (blue) and ruby (red). It also occurs in some igneous and metamorphic rocks. The primary use for corundum is in grinding wheels, discs, and other abrasive agents that utilise its hardness.

cosecant the function of an angle in a right-angled triangle given as the reciprocal of the sine function. The cosecant of angle A is 1/sine A.

cosine a trigonometric function used to determine the value of a specific angle in a right-angled triangle by calculating the adjacent-to-hypotenuse ratio (adjacent, hypotenuse refers to the sides of the triangle).

cosmic rays "rays" composed of ionizing radiation from space and comprising mainly PROTONS, ALPHA PARTICLES (helium nuclei), and a small proportion of heavier atomic nuclei. When these particles interact with the Earth's atmosphere, secondary radiation is created, including MESONS, ELECTRONS, POSITRONS and NEUTRONS. Cosmic rays seem to emanate from three sources: galactic rays, possibly formed by supernovae explosions; solar cosmic rays; and the SOLAR WIND.

cosmology the study of the universe, including origin, structure and evolution. There are two famous hypotheses—the Big Bang, which superseded the other, the Steady-State Theory.

coulomb a unit of electric charge (C): the amount of electricity carried through a given region in one second by a current of one AMPERE (A), i.e.

Charge (C) = Current (A) x Time (s).

covalent bond the joining of two atoms due to the equal sharing of their electrons. In a single covalent bond, one electron pair is equally shared between two atoms (as in the hydrogen molecule, H_2). In a double bond, two electron pairs are shared, and in a triple bond, three electron pairs are shared between two atoms. Covalent bonds are usually formed between non-metallic elements in which the atoms have a strong but equal attraction for electrons, resulting in the formation of strong, stable bonds.

CPU (*abbreviation for* central processing unit) in computing, the electronic device that accepts and processes information that can be modified and stored and then output.

Crab Nebula a NEBULA that is the remains of a supernova in AD 1054. It is an expanding source of radio waves and X-rays. At the centre is a PULSAR, which is waning in its emissions.

cracking a process much used in industry where large, complex molecules, e.g. HYDROCARBONS, are broken down into smaller molecules. Heat is usually the primary agent but pressure and CATALYSTS are also used. It is used particularly in the PETROLEUM industry.

craton a stable area of the Earth's crust, consisting of undeformed sediments over older crystalline BASEMENT. Tectonism is very limited in its effect, and only minor volcanic activity occurs. A typical example is the Canadian Shield.

crepuscular rays (of twilight) rays of sunlight, emanating from the Sun when below the horizon, which are rendered visible by impinging on hazy cloud or by shining through gaps in cloud (to give the effect called Jacob's Ladder).

cresols *or* **hydroxytoluenes** ($CH_3C_6H_4OH$) are important in the production of resins, pesticides and herbicides, disinfectants and ANTIOXIDANTS, and also dyes. Cresols occur in COAL TAR and cracked (*see* CRACKING) NAPHTHA.

cross-bedding a primary SEDIMENTARY STRUCTURE formed by the gradual movement of the slip face in subaqueous or subaerial sand dunes, waves, or bars. In ARENACEOUS rocks this produces inclined planes within the bedding (*see* BEDDING PLANE) of the rock, relating to the direction of the original current.

crossing-over the reciprocal exchange of genetic material between two HOMOLOGOUS CHROMOSOMES during MEIOSIS. This process occurs at a site called the chiasma of the two chromosomes and is responsible for GENETIC RECOMBINATION.

crucible a heat-resisting vessel used in laboratories and industry, in which metals, rocks and other materials are melted, fused or reacted. Crucibles may be made of refractories, nickel, platinum, etc.

crude oil PETROLEUM in its unrefined state.

crust *see* **continental drift**.

cryogenics the study of materials and their behaviour at very low temperatures. It is usually dependent upon the use of liquefied gases, known as **cryogens**.

crystal an ordered homogeneous solid bounded by faces that are usually flat, and with a limited chemical composition and specific atomic structure (*see* LATTICE). Examination of crystals, even with the naked eye, shows a regularity in the position of faces and edges. This is part of a crystal's SYMMETRY, which is defined by three parameters: planes of symmetry; axes of symmetry; and a CENTRE OF SYMMETRY. Each crystal type can be defined using these factors. A plane of symmetry divides a crystal into like halves in similar positions (to create mirror images); for example, a cube has nine planes that divide it into two halves. An axis of symmetry is when a crystal, on being rotated, occupies the same spatial position more than once in a complete turn. Depending upon the extent of the symmetry of a crystal, it may occupy the same position 2, 3, 4 or 6 times in one rotation; the cube possesses three axes of 4-fold symmetry, four of 3-fold symmetry, and six of 2-fold symmetry. Finally, a crystal may have a centre of symmetry when like faces and edges are arranged in pairs in like positions on opposite sides of a central point. Other features that apply to crystals include HABIT and TWINNING.

cube *see* **polyhedron**.

cubic equation an algebraic equation of standard form: $x^3 + 3ax + b = 0$. The solution of such equations is complex and they have little practical value.

cumulonimbus large, bulging clouds that reach great heights, with the upper reaches forming anvil shapes or plumes. The base is dark, usually producing precipitation, often with VIRGA.

cumulus well-defined cloud in the form of separate masses comprising bulges and towers, with a darker, often flat, base.

cupola a form of intrusion rising from, and much smaller than, a larger intrusion (e.g. a BATHOLITH) and shaped like a dome.

curing the setting of a hot liquid thermosetting PLASTIC, which usually occurs during the moulding operation.

current the flow of charge in an electrical CIRCUIT. Current (I) is measured in AMPERES (A).

cybernetics a science developed by the American mathematician Norbert Wiener (1894-1964). Cybernetics is the comparative study of organization, regulation and communication within control systems. Such studies can provide very useful information concerning the role of man and machine at work, e.g. reducing human error or devising a computer capable of making managerial decisions.

cyclic compound *see* **closed-chain**.

cyclone *see* **depression**.

cyclothem the result of cyclic or rhythmic sedimentation, whereby a set of deposits is repeated in a sequence. It occurs typically in coal-bearing rocks.

cyclotron a machine that produces high-speed charged particles for use in nuclear experiments. The particles are accelerated by motion along a spiral path between D-shaped electrodes between a magnet, in a vacuum. Energies of several million electron volts are imparted to the particles, which ultimately are deflected, electrically, out of the cyclotron.

cytogenesis formation and development of cells. **Cytogenetics** is the associated study, involving cell structure and the structure and behaviour of chromosomes in relation to inheritance.

cytology the study of cells, their structure and function. Microscopy, particularly ELECTRON MICROSCOPY, has permitted the study of cell components, including the NUCLEUS and CHROMOSOMES.

cytoplasm the part of the cell outside the NUCLEUS but within the cell wall.

cytosine ($C_4H_5N_3O$) a nitrogenous base component of the NUCLEIC ACIDS, DNA and RNA, which has a pyrimidine structure and BASE PAIRS with GUANINE.

D

Dalton's atomic theory although superseded by subsequent experimentation and theory, this thesis formed one of the fundamental stages in the development of chemistry. In the theory, the English chemist and physicist John Dalton (1766–1844) stated that matter consists of indivisible discrete particles (atoms), which are identical for one element, and that chemical action occurs through attraction between atoms.

Dalton's law of partial pressures in a mixture of gases, the total pressure exerted is the sum of the partial pressures of those gases present. The partial pressure is the pressure a gas in a mixture would exert if it alone occupied the same volume.

dark nebulae clouds of obscuring gas and dust common throughout the MILKY WAY and seen in other galaxies.

dark reaction the stage in PHOTOSYNTHESIS that occurs in the stroma of CHLOROPLASTS in plants and is not directly dependent on light (*see* CALVIN CYCLE).

Darwinism the theory of evolution described by the British naturalist Charles Robert Darwin (1809-1882). He first established the evolutionary theory of environmental forces acting as agents of natural selection on successive generations of organisms. The theory had five elements:

Darwinism

(1) Individuals have random variability, more so if they sexually reproduce.

(2) Reproductive capacity inevitably leads to competition both within (intra) and between (inter) species as populations tend to remain a set size (if there are no major external influences to upset this, e.g. mass-slaughter of animals by humans or an environmental catastrophe such as a huge oil slick).

(3) Some individuals are better adapted than others to their environment, which will help in their survival and reproductive success, i.e. fitness.

(4) Some of the characteristics that make parents successful in their ability to survive and reproduce are inherited by their offspring, thus increasing the probability of success for those offspring. These characteristics become more widespread in the population. This is evolutionary change; the surviving species has adapted better to an environmental niche, i.e. survival of the fittest.

(5) The descendants of a single stock tend to diverge and become adapted to many different environmental NICHES. Darwin's paper, "Origin of Species" (1859), was greatly criticized by both the religious and scientific establishments of that period. He did not know of GENES as the units of inheritance, so it was understandable that he believed in the inheritance of acquired characteristics, as this would be a rational explanation for the phenomena he observed. With the knowledge that information regarding an organism's PHENOTYPE is present in their genes, then Darwin's theory can be summarized as follows—if the organisms have the capacity for genetic variability, then new species can emerge that will adapt to new environments, while the old species, which

are no longer suited to the surrounding environment, will eventually die out.

DC *see* **direct current**.

Debye-Hückel theory the activity coefficient of an ELECTROLYTE (i.e. the ratio of activity to the true concentration of a substance or the *effective* concentration) depends greatly upon concentration in solution. Due to forces of attraction, weak electrolytes have ions that are surrounded by ions of the opposite charge, thus reducing their activity. The Dutch-American physicist Peter Debye (1884–1966) and the German physicist Erich Hückel (1896–1980) explained abnormal coefficients in this manner for weak electrolytes.

decay the process whereby a radioactive (*see* RADIOACTIVITY) substance is transformed into its "daughter" products, and ultimately may form a stable atom, (*see also* URANIUM).

decibel a unit for measuring the intensity of sound levels; it has the symbol dB.

decimal a structured system of numbers based on 10. *See also* PLACE VALUE NOTATION, and APPENDIX 4.

declination the angle between magnetic north and geographic (true) north. Also in astronomy, the angular distance of a heavenly body from the CELESTIAL EQUATOR.

declination circle a graduated circle on an EQUATORIAL telescope that allows the telescope to be set at a specific declination.

decomposer any organism that breaks down dead organic matter, whether organisms or plant and animal waste, to obtain energy, leaving simple organic or inorganic material. Earthworms, but mainly bacteria and fungi, fulfil this role. Carbon dioxide is released and the accompanying heat is useful, where conditions permit, in killing parasitic and pathogenic bacteria, eggs etc (*see also* CARBON CYCLE, NITROGEN CYCLE).

decomposition the breakdown of a substance into simpler molecules or atoms by various means, e.g. pyrolysis (heat), ELECTROLYSIS, or biological agents (biodegradation).

defect a break or discontinuity in the structural arrangement of a crystal, whether in the pattern of ELECTRONS, ATOMS or IONS. The discontinuity may take the form of a vacancy or point defect, or be a linear break (line defect) or dislocation.

deformation processes that rocks undergo, producing stress and geometrical changes that constitute the strain, or deformation. Included are FOLDing, faulting (*see* FAULT), emplacement of igneous INTRUSIONS and gravity-controlled structures.

degeneration in evolutionary terms, returning from a complex to a less complex state. More specifically, reduction or loss of organs in the course of evolution; or change in cells, tissues or organs due to disease, resulting in a loss of function and possibly breakdown.

degradation the breakdown and simplification of complex molecules. In physics, loss of KINETIC ENERGY due to collision.

degrees of freedom the lowest number of variables defining the state of a system, e.g. temperature and pressure of a gas.

dehydration in chemistry, the removal of water that is chemically held in a molecule or compound by the action of heat, often with a CATALYST or chemical acting as a dehydrating agent. SULPHURIC

ACID is an example of the latter. In medicine, dehydration is the excessive, often dangerous, loss of water from the body tissues.

dehydrogenase a class of ENZYME that facilitates the removal of hydrogen in biological reactions. They occur in numerous biochemical pathways but are particularly important in AEROBIC RESPIRATION.

deliquescence when a substance picks up moisture from the air and may eventually liquefy. It is due to the substance having a very low water VAPOUR PRESSURE, lower than that of the surrounding air, thus water is absorbed.

delta a roughly triangular area of sediment formed at the mouth of a river. It is due to a current laden with sediment entering a body of water, resulting in a reduction of the current's velocity and thus its carrying capacity. Much of the sediment is therefore deposited on entering the lake or sea. The shape of the delta depends upon various factors including water discharge, climate, tides and sediment loads. Modern examples include the Mississippi and the Nile.

denaturation the disruption of the weak bonds that hold a PROTEIN together, usually caused by extreme heat, or addition of a strong acid or alkali. Most denatured proteins precipitate and cannot refold into their original structure. During cooking, the proteins within an egg become denatured, causing both the yolk and the white of the egg to solidify. Extremes of temperature are fatal to the majority of all animals when the protein molecules, called ENZYMES, undergo irreversible denaturation and cannot perform their essential function as CATALYSTS of the biochemical processes necessary for life.

dendrochronology the science of dating through the study of annual tree rings.

denominator the number below the line in a VULGAR FRACTION e.g. 3 in $^2/_3$.

density the mass per unit volume of a substance. It is measured in units of kilograms per cubic metre (kgm^{-3}) and can be calculated using the following equation:

$$\text{density } (d) = \text{mass } (M) / \text{volume } (V)$$

Thus, the density of a constant volume of gas will increase if the gas is transferred to a container smaller than the initial one, but the density of the same volume of gas will decrease if the gas is transferred to a bigger container.

denudation the "stripping bare" of a land surface, involving WEATHERING and EROSION. In theory, the end point would be the removal of hills to the existing datum. A process complementary to sedimentation.

deoxyribonucleic acid *see* **DNA**.

dependent variable in a mathematical expression, the quantity with a value determined by the other INDEPENDENT VARIABLES. For example, in the equation $y = 6x + 3$, y is the dependent variable, i.e. the value of y is dependent on the value inserted for x (the independent variable).

deposition the laying down of sediments (sedimentation). It also applies to mineral veins.

depression an area of low pressure, also called a cyclone, with particular patterns of wind flow—anti-clockwise in the northern hemisphere and clockwise in the southern hemisphere. Often associated with unsettled, stormy weather.

derivative the rate of change of a FUNCTION relative to a specific value given to the INDEPENDENT VARIABLE.

The derivative of a function is obtained by DIFFEREN-
TIATION and usually refers to the gradient of a graph.
For example, the graph of the function $y = 3x^2$ has
a derivative of $6x$ and will therefore have a gradient
of 12 when $x = 2$.

desalination the production of fresh water from sea
water. A number of processes may be used, including EVAPORATION, DISTILLATION, freezing, and reverse
OSMOSIS, although these tend to be costly.

desiccant a substance that absorbs moisture (*see*
HYGROSCOPIC) and is thus used as a drying agent.
Examples are calcium chloride, silica gel, and phos-
phorus pentoxide.

detergent a soluble substance that acts as a cleaning agent and is particularly effective in the removal of grease and oils (both HYDROCARBONS). The
hydrophobic part of a detergent molecule will readily interact with hydrocarbons, whereas the
hydrophilic part will readily interact with water
molecules. When added to an immiscible mixture of
water and, say, grease, the "dual reactive" nature of
the detergent brings the water and grease together,
thus allowing them to be rinsed away.

detrital mineral a mineral derived from a parent
rock by the mechanical breakdown of the rock by
WEATHERING and EROSION. Diamond, zircon and gold
are typically resistant minerals.

deuterium an ISOTOPE of the atom hydrogen that
constitutes 0.015 per cent of naturally occurring
hydrogen. It is indicated by either the symbol D or
$_1^2H$, and will react with oxygen to form heavy water,
i.e. D_2O.

devitrification a change in glass from the glassy
state, to a crystalline one, due to the formation of

minute crystals. The same applies in geology to glassy igneous rocks (e.g. OBSIDIAN) that become dull, finely crystalline rocks. However, their origin can be determined by the presence of textures typical of glasses.

dew condensation of water vapour in the air, producing water or moisture, due to the temperature falling causing the vapour to reach saturation. Surfaces become cooled by radiation, and when the temperature goes below the DEWPOINT, condensation occurs from the air in contact with the cool surfaces.

dewpoint the temperature at which air becomes saturated with water vapour and deposits dew.

dextrorotatory compound any substance that, when in crystal or solution form, has an optically active property that rotates the plane of polarized light to the right (clockwise), e.g. the sugar D-glucose (D stands for dextrorotatory).

diachronism when a geological formation exhibits varying ages in different locations due perhaps to a marine TRANSGRESSION.

diagenesis the changes that occur in a sediment after DEPOSITION, at low pressures and temperatures. These processes include compaction of the grains, DISSOLUTION, replacement, cementation (i.e. cementing together of grains, particles or fragments by materials deposited from fluid percolating through the pore spaces of the rock), and sometimes recrystallization. By these means, for example, a sand becomes a sandstone.

dialysis a method for separating small molecules from the larger ones present in a solution. Dialysis occurs in the kidneys of all vertebrates to remove the waste products of METABOLISM.

diapir an intrusion, dome-like in shape, that pushes upwards into existing, denser rocks. In so doing, it domes up the cover rocks. This occurs with granite and salt masses, and the process is called **diapirism**.

diastase(s) alternative name for amylases, ENZYMES that break down starch. Plants contain β- and α-amylase, whereas animals contain only the latter. Specifically, diastase is the part of malt containing β- amylase, and is thus important in brewing.

diastrophism an old term used to describe the large-scale deformation of the Earth's crust, including FOLDing, FAULTing, and plate movement.

diazo compounds a group of organic compounds that may form an AZO group, i.e. 2 nitrogen atoms (-N = N-) where one is attached to a carbon atom. When attached to certain aromatic molecules, the resulting compounds are important as drugs and dyes.

dibasic acid a substance containing two replaceable hydrogen atoms per molecule. A dibasic acid will react with a BASE to form either an acidic salt (only one hydrogen atom is replaced) or a normal salt (both hydrogen atoms are replaced). Thus the products of the neutralization of a dibasic acid depend upon the quantities of the acidic and basic reactants.

dichotomy dividing into two equal parts, e.g. in botany, where the growing point of a plant divides to produce two further, equal, growing points that similarly divide after more growth. In astronomy, when a planet is seen half-illuminated, e.g. the Moon.

dicotyledon *see* **monocotyledon**.

dielectric a material, gas, liquid or solid that does

not conduct an electric current—an INSULATOR. Dielectrics can be used for capacitors (*see* CAPACITANCE), terminals, and cables.

differentiation a procedure used in CALCULUS for finding the derivative of a function. There are various methods of differentiation, depending on the simplicity or complexity of the function and its corresponding graph. One of the simplest forms of differentiation is for the common function, $f(x) = x^n$, which has the derivative $f'(x) = nx^{n-1}$. For example, the function $f(x) = 3x^3$ has the derivative $f'(x) = 9x^2$, whereas the function $y = 2x^4 + 6x^2$ has the derivative $dy = 8x^3 + 12x$. For more complex methods of differentiation *see* CHAIN RULE, PRODUCT RULE, QUOTIENT RULE.

In geology (as **magmatic differentiation**) the process of several rock types being produced from one parental MAGMA. It involves the crystallization of certain crystals, which are in effect removed from the magma, thus altering the composition, resulting in the formation of a group of different minerals.

diffraction the bending of waves round an obstacle such as the straight edge of a barrier. The diffraction of all waves—water, light, sound and electromagnetic—around a suitable object can be detected by a change in the wavefront shape. There will be no change in the wavelength or wave frequency, and consequently the speed of the wave will be constant provided the properties of the surrounding medium remain constant.

diffusion the natural process by which molecules will disperse evenly throughout a particular substance. The molecules will always travel down their concentration gradient; that is, they will move from a region where they are highly concentrated to a region where they are lower in concentration within that substance. Diffusion occurs in gases, liquids and, depending on the size of the molecules, across the membranes of cells and ORGANELLES within cells.

diffusivity the rate at which heat diffuses through a material, measured as $m^2\ s^{-1}$.

digital a term signifying the use of a numerical code as opposed to the mechanical indicators on a dial. In a digital clock, illuminated numerical figures are used to display time, whereas a conventional clock has indicators that move around a dial.

dimer a molecule formed by two identical molecules bonding together. The formation of THYMINE-thymine dimers in the same strand of DNA occurs in cells damaged by exposure to ultraviolet light.

diode a device that will allow current to flow in only one direction. A diode valve has a solid surface that, when heated, will emit electrons and hence become the negatively charged plate (CATHODE). The current will travel towards a positively charged plate, the ANODE, which is connected to the positive terminal of a power source such as a battery. Overall, the arrangement of the cathode, anode and heating supply in a vacuum is called a diode valve or vacuum tube.

dip the angle of inclination of a plane, measured from the horizontal and perpendicular to the STRIKE.

diploid a term used to describe a cell having two of each CHROMOSOME in its nucleus. All diploid organisms will have HOMOLOGOUS pairs of chromosomes, with each member having a similar distinctive shape. The homologous chromosomes of a pair contain GENES that code for the same products during PROTEIN synthesis, although the information could result in a different form of protein and thus a characteristic (*see* ALLELE). Humans are diploid, with each cell of the body (except GAMETES) having 46 chromosomes; that is, 22 pairs of homologous AUTOSOMES and a pair of sex chromosomes.

dipole one of two equal and opposite charges (electric dipole) or magnetic poles (magnetic dipole), separated by a short distance. In an ATOM, a transient electric dipole is generated by the random distribution of its continuously moving electrons, and in a MOLECULE a dipole is the result of unequal sharing of electrons within a bond. If two atoms of a molecule differ in ELECTRONEGATIVITY, one atom will have a greater attraction for the electrons of the bond than the other. This will cause one end of the molecule to be slightly positive and the other end to be slightly negative. For example, water molecules (H_2O) are dipoles, as the greater electronegativity of the oxygen atom (3.5) will have a stronger attraction for the electrons of the bonds than does the electronegativity of the hydrogen atoms (2.1). Thus, the bonding electrons spend more time around the oxygen atom, giving the oxygen end a slightly negative charge and the hydrogen end of the H_2O molecule a slightly positive charge.

direct current (DC) an electric current flowing in one direction only.

disconformity an UNCONFORMITY in which the beds above and below the erosion surface are parallel.

discriminant *see* **quadratic equation**.

disintegration the process, especially but not only radioactive DECAY, where one or more particles is ejected from a NUCLEUS.

displacement the calculation determining the

change of position of an object by relating its final position to its initial position. Displacement is always independent of the path taken and is measured in metres (m).

dissociation the process by which a compound breaks up into smaller MOLECULES, or IONS and ATOMS.

dissolution the dissolving of a substance in a liquid to form a homogeneous solution.

distance in physics, the measurement of how far an object has travelled along a specific path. In mathematics, distance is the length of the line needed to join particular points and is similar to displacement. The units of distance are metres.

distillation the separation of a liquid solution into its various components by initially heating the liquid into vapour and then cooling the vapour so that it condenses and can be collected. The different components of a solution can be collected at different intervals as each component will have a unique BOILING POINT.

distribution law the ratio of the concentrations of the SOLUTES dissolved in two immiscible liquids, which are in equilibrium with each other. The value is constant when the temperature is also constant. The temperature must be constant as heat increases the solubilities of liquids in liquids.

diurnal daily—recurring every 24 hours (*see* CIRCADIAN).

diurnal parallax the effect caused by the Earth's rotation such that to an observer, it *appears* that the position of a body in the solar system changes.

diurnal range the daily changes within one component of meterology, e.g. temperature.

divergent junction *or* **constructive boundary** a

boundary between plates that are moving apart because new crustal material is being created. Features associated with these boundaries include EARTH-QUAKES and volcanism (*see also* MID-OCEANIC RIDGE).

dizygote a term used to describe twins who have developed from two separate ova and therefore have different genetic characters (*see also* FRATERNAL TWINS).

DNA (*abbreviation for* **deoxyribonucleic acid**) a NUCLEIC ACID and the main constituent of chromosomes. Made up of a double helix of two long chains of linked NUCLEOTIDES, BASE PAIRed between ADENINE and THYMINE or CYTOSINE and GUANINE, it transmits genetic information in the form of GENES from parents to offspring.

DNA polymerase an ENZYME that helps in the replication of new DNA strands by using the uncoiled double-helix of DNA as a template on which to add free NUCLEOTIDES. Three DNA polymerases have been identified, and all of these enzymes help catalyse the step-by-step addition of a DNA nucleotide to the end of a growing DNA chain:

DNA Polymerase I—gap filling and repair enzyme
DNA Polymerase II—function unknown at present
DNA Polymerase III—responsible for the majority of DNA synthesis

Dobson spectrophotometer an instrument used in meteorology to measure atmospheric OZONE and with which it is possible to calculate the vertical distribution of ozone.

doldrums relatively calm ocean areas, around the equator, which have low pressure and light winds.

dolerite a dark, medium-grained IGNEOUS rock occurring commonly as DYKES, SILLS or volcanic PLUGS. BASALT is its fine-grained equivalent. Dolerites contain a plagioclase variety of FELDSPAR and the PYROXENE augite. OLIVINE may also be present, and other minerals present in small amounts include magnetite (Fe_3O_4), and ilmenite ($FeTiO_3$).

dolomite a widespread rock-forming mineral with composition $CaMg(CO_3)_2$. Also a sedimentary rock that is usually formed by dolomitization of LIMESTONE, i.e. the addition of magnesium to the sediment or rock at some point after DEPOSITION, possibly during DIAGENESIS. Dolomitization and recrystallization in the late stages of diagenesis increase the rock's POROSITY, making dolomites important petroleum RESERVOIR rocks.

dominance a genetic concept that a certain ALLELE of a gene will mask the expression of another allele known as the recessive. The PHENOTYPE of the individual is a result of the expression of the dominant allele with the recessive allele (*see* HETEROZYGOTE, HOMOZYGOTE) remaining undetected as it will have no effect on the phenotype.

doppler effect a change in the observed frequency of a wave as a result of the relative motion between the wave source and the detector. The doppler effect is responsible for an ambulance siren having a higher pitch as it approaches you and a lower pitch as it recedes into the distance.

double bond the linking of two atoms through two COVALENT BONDS in a compound, i.e. the sharing of two pairs of electrons.

double decomposition a reaction between two substances that results in their breakdown and the subsequent formation of two different substances comprising their parts, e.g.

$$AB + XY \longrightarrow AX + BY$$

101

downthrow the downward displacement of one side of a FAULT, relative to the other.

drift a general term (now often superseded) used for superficial deposits, i.e. as opposed to solid rock formations. More specifically, it is a sediment deposited by glacial ice.

drosometer an instrument to measure dew.

drowned valleys river valleys that have been submerged due to a rise in sea level, often caused by post-glacial melting.

drumlin an oval accumulation of BOULDER CLAY that forms a small hill. Typically, it has one end blunt, the other pointed because of streamlining by the movement of the ice from which they were deposited. They are commonly found in groups, with their line of elongation reflecting the movement direction of the ice.

dry ice the solid form of carbon dioxide (CO_2), which is used as a refrigerant. It sublimes (*see* SUBLIMATION) at $-78°C$ without forming a liquid, hence the term "dry" ice.

dry valley as the name suggests, a valley feature originally created by water EROSION, but now dry. This may be due to a fall in the WATER TABLE, removal of earlier PERMAFROST conditions, or to "capture" of the flow by another river or stream.

ductility the property of metals or alloys that enables them to be drawn out into a wire and to retain strength when their shape is changed.

dune an accumulation and movement of sediment, usually sand, by the wind. The size varies, but in the Sahara Desert can rise to 300 metres (*c.*1000 ft) in height. There are three types of dune, depending upon morphology and formation.

duodenum the first part of the small intestine of vertebrates, which connects the stomach to the ileum and into which secretions from the gall bladder and pancreas are emptied.

dyke igneous INTRUSION that occurs as a sheet-like body with near-parallel sides. Dykes are normally discordant, cutting across the country (host) rock, and are usually vertical or nearly so.

dyke swarm the occurrence of numerous DYKES in a radial, sub-parallel or some other pattern.

dynamics the branch of physics that deals with the motion of objects under the action of the forces responsible for changes in the motion of those objects.

dynamo a machine that generates electric currents using a rotating coil as a conductor and powerful magnets to create a MAGNETIC FIELD.

dyne the unit of force prior to the use of NEWTON in the SI notation (*see* SI UNITS). The dyne is the force required to produce an acceleration of 1cm s^{-2} in a mass of 1g. 1 Newton = 10^5 dynes.

E

Earth (*see* APPENDIX 6 for physical data) the third
planet in the SOLAR SYSTEM with its orbit between
Venus and Mars. It is a sphere, flattened at the
poles, with an equatorial radius of 6378 kilometres
(3963 miles). It comprises a gaseous ATMOSPHERE, a
liquid HYDROSPHERE and solid LITHOSPHERE. The core
is part liquid and has a relative density of 13, and a
temperature of almost 6500K. It is composed of
nickel-iron separated into an inner and outer core.

earthquake the often violent movement of the earth
due to tectonic upheaval caused by the sudden
release of accumulated stress, perhaps along a fault
or fault zone. Earthquakes are classified by depth of
focus: shallow—less than 70 km (43.5 miles); intermediate—70–300 km (43.5–186.4 miles); deep—
more than 300 km (*see* EPICENTRE and RICHTER SCALE).
Attempts to predict earthquakes rely upon measurement of stress increases, but can so far be related
only to an increasing risk of activity.

ECG (*abbreviation for* **electrocardiograph**) equipment used to record the current and voltage associated with contractions of the heart.

eclipse the total or partial disappearance of one
astronomical body by passing into the shadow of
another. A solar eclipse occurs when the new moon

passes between the earth and the sun. A lunar eclipse occurs when the moon moves into the shadow of the earth, i.e. the earth is situated between the sun and the moon.

ecology the study of the relationship between plants and animals and their environment. Ecology is also known as bionomics and is concerned with, for example, predator-prey relationships, population dynamics and competition between species.

ecotype a population that is genetically adapted to the local conditions of its particular habitat.

ecosystem an ecological community that includes all organisms which occur naturally within a specific area.

EDTA (ethylene diamine tetra-acetic acid $CH_2N(CH_2COOH)_2$ a chelating agent (*see* CHELATION) often used in TITRATION of metal ions. Because of its readiness to combine with metals, it can be used to protect enzymes from inhibition by metals i.e. the prevention of normal enzyme activity because the inhibitor binds to the active site on the enzyme, thus blocking the substrate. EDTA can also be used with some culture media as a slow-release agent for metal ions.

efflorescence the condition when a salt hydrate loses water in air, forming a powder on its surface.

Einstein, Albert (1879-1955) a German-born American physicist who formulated the RELATIVITY theories and carried out important investigations in THERMODYNAMICS and radiation physics. In 1921, he received the Nobel Prize in physics. He became a professor of mathematics at Princeton University, New Jersey, in 1933 after fleeing Nazi Germany.

elastic limit (*see* HOOKE'S LAW) the point beyond

which a body will deform when stress is applied, rather than returning to its original state.

elasticity a property of any material that will stretch when forces are applied to it and recover when the forces are relaxed. To stretch a spring or any other elastic material, equal but opposite (direction) forces must be applied to two areas of the material. All materials are elastic to some extent and will obey HOOKE'S LAW if the forces applied do not cause permanent deformation.

elastomer a material, usually synthetic, that has properties similar to natural rubber, e.g. a capacity to revert to its original state after extension. In addition to rubber, elastomers include synthetic rubbers (polybutadienes) and other materials, such as butadiene COPOLYMERS.

electric current a flow of electric charge through a CONDUCTOR. The charge may be carried by means of electrons, ions or holes. Hole is the term for the absence of an electron in the valence structure (*see* VALENCY) of a body. The movement of an electron in to the hole creates new holes and therefore "conduction by holes" (*see also* CURRENT).

electric field the invisible force that always surrounds any charged particle. When one charged particle is within proximity of the electric field of another, each will feel and exert a force. If the charged particles are opposite (unlike), then they will attract since opposite charges attract. However, two particles with the same charge, i.e. both positive or both negative, will repel each other. An electric field is diagrammatically represented by lines along which a free positive charge would theoretically move, and so the arrows will always point towards a negative charge and away from a positive charge. The strength of an electric field is represented by the density of the drawn lines, but can be measured in either volts per metre (Vm^{-1}) or newtons per coulomb (NC^{-1}).

electric storm an atmospheric disturbance in which the air is charged with static electricity that, with clouds, produces thunderstorms.

electrocardiograph *see* **ECG**.

electrode a conductor which facilitates passage of an electric current into or out of a liquid (as in ELECTROLYSIS) or a gas (gas tube) or a vacuum (valve). *See also* CATHODE, ANODE and CONDUCTION.

electrodeposition deposition by electrolysis of a substance on an electrode. It is used particularly to deposit one metal onto another, as in electroplating where the cathode is the object to be coated, or in electroforming (the production of metal items by metal deposition upon an electrode by electrolysis).

electrolysis chemical decomposition achieved by passing an electric CURRENT through a substance in solution or molten form. IONS are created, which move to electrodes of opposite charge where they are liberated or undergo reaction.

electrolyte a compound that dissolves in water to produce a SOLUTION that can conduct an electrical charge. The electrical conductivity of the solution is a result of the compound undergoing ionization to form mobile IONS. If the substance is completely ionized, it is termed a strong electrolyte, e.g. any strong acid such as sulphuric acid, but if it is only partially ionized it is termed a weak electrolyte, e.g. any weak acid such as ethanoic (acetic) acid.

electromagnet an electromagnet is created when

107

an electric current flows through a coil surrounding a soft iron core, and the latter becomes a magnet.

electromagnetic induction the production of an ELECTRIC CURRENT when a conductor is moved through a MAGNETIC FIELD. A current is induced to flow only while there is a changing magnetic field due to the movement of the conductor or magnet. The direction of the induced current flow depends on the pole orientation of the magnet, i.e. north-south or south-north, and whether the magnet is moving towards or away from the conductor. The magnitude of an induced current is dependent on the strength of the magnetic field, the cross-sectional area of the conductor, and both the speed and direction of the relative motion between the conductor and the magnetic field.

electromagnetic waves the effects of oscillating electric and magnetic fields that are capable of travelling across space, i.e. they do not require a medium through which to be transmitted. The spectrum of electromagnetic waves is divided into the following categories:

	wavelength (m)	*frequency* (Hz)
gamma rays	10^{-10}—10^{-12}	10^{21}—......
X-rays	10^{-12}—10^{-9}	10^{21}—10^{17}
ultraviolet radiation	10^{-10}—10^{-7}	10^{18}—10^{15}
visible light	10^{-7}—10^{-6}	10^{15}—10^{14}
infrared radiation	10^{-6}—10^{-2}	10^{14}—10^{11}
microwaves	10^{-3}—10	10^{11}—10^{7}
radiowaves	10 — 10^{6}	10^{7}—10^{2}

All electromagnetic waves travel through free space at a speed of approximately 3×10^8 ms^{-1}, known as the SPEED OF LIGHT. The only electromagnetic waves that are readily detected by the eye are visible light waves. These consist of various wavelengths, which correspond to the colours red, orange, yellow, green, blue, indigo and violet. The colour red has the longest wavelength, lowest frequency, and the colour violet has the shortest wavelength, highest frequency. An object that appears to be red in colour, for example, has absorbed all the light waves from the blue end of the spectrum while reflecting the ones from the red end.

electron an indivisible particle that is negatively charged and free to orbit the positively charged NUCLEUS of every atom. In the traditional model, electrons move around in concentric shells. However, the latest concept, based on quantum mechanics, regards the electron moving around the nucleus in clouds that can assume various shapes, such as a dumb-bell (two electrons moving) or clover leaf (four moving electrons). The shape and density of the outermost electronic shell will help determine what reactions are possible between particular atoms and molecules, e.g. whether an atom will easily gain or lose electrons to form an ION.

electron affinity the energy change associated with the capture of an electron by a single, gaseous atom (*compare* ELECTRONEGATIVITY). The process is EXOTHERMIC when a single electron is captured by a neutral atom (first electron affinity), but any further electron captures (second or higher electron affinities) can be ENDOTHERMIC processes.

electronegativity (*plural* **electronegativities**) a measure of the power of an atom within a MOLECULE to attract electrons. Every element in the PERIODIC TABLE is given an electronegativity rating based on an arbitrary scale in which fluorine is given the highest rating of 4, as it has the most electronegative atoms. Electronegativity differences between the different atoms within a molecule can be used to estimate the nature of the bonds formed between those atoms, i.e. whether it is a COVALENT, IONIC or POLAR COVALENT BOND.

electron microscope a microscope that uses a beam of electrons rather than a beam of light impinging upon the object. This gives much greater resolution. In the *transmission* electron microscope, the electron beam passes through a very thin slice of the object, and an image is created due to the scattering of the beam, which is focused and enlarged dramatically onto a fluorescent screen. *Scanning* electron microscopy involves scanning the surface of a sample, and the image is generated by secondary electrons. The magnification of the object is less with scanning electron microscopy, but a three-dimensional image is created.

electron pair two electrons from the outer shells of two atoms which are shared by the adjacent nuclei to form a bond.

electron-probe (micro-analysis) a means of analysing a very small sample of a substance. A very finely focused beam of electrons strikes the sample and creates excitation, resulting in the emission of X-rays that identify the ELEMENTS present by their characteristic wavelength. The method provides both a QUANTITATIVE and QUALITATIVE determination.

electron spin resonance (ESR) a phenomenon utilized in ESR spectroscopy and used to determine molecular structure, and to gather information about chemical BONDS, free RADICALS and crystal impurities. The technique relies upon the spin of unpaired electrons (*see* QUANTUM NUMBERS), which upon the application of a magnetic field align in one of two ways, each at different energy levels. Transitions (in the microwave region of the spectrum) between these levels can be initiated and the information used to define the structures observed.

electrophile a molecule that will readily accept electrons from another molecule. An electrophile usually has an electron-poor site as its functional group, as in the ion, NO_2^+, or the acid, HBr, and will thus attack the high electron density region of other molecules, such as the double bond in ALKENES.

electrophoresis a method for separating the molecules within a solution using an electric field. The molecules will move at a speed determined by the ratio of their charge to their mass. All electrophoretic separations involve placing the mixture to be separated onto a porous, supporting medium, such as filter paper, or a plant-derived gel called agarose that has been soaked in a suitable BUFFER. An electric field is applied, causing the molecules, now dissolved in the conducting buffer solution, to move at a rate according to their charge-to-mass ratio. The various molecules can be identified by comparing their final position on the supporting medium with the position of known standards. Gel electrophoresis uses a synthetic polymer, such as polyacrylamide, as the supporting medium, and can be used to separate the different lengths of DNA chains found within the cellular nucleus of an individual. This makes gel electrophoresis an essential technique of "genetic fingerprinting," as the separated DNA strands will contain genes unique to a particular person, thus helping in identification.

electrostatics the part of the science of electricity dealing with the phenomena associated with electrical charges at rest. Any material containing an electric charge that is unable to move from atom to atom is called an INSULATOR (as opposed to a CONDUCTOR, which allows elecric charge to flow throughout it). It is possible to negatively charge an insulator, such as polythene, by transferring electrons to it using a woollen cloth. A similar effect is sometimes produced when a plastic comb is used to comb dry hair.

element a pure substance that is incapable of separating into a simpler, different substance when subjected to ordinary chemical reactions. There are 105 elements known to us, but only 93 of these occur naturally, and the others have been created in laboratories. The elements are classified into the PERIODIC TABLE, according to the number of protons in their nucleus, i.e. their atomic number. Each element of the periodic table consists of atoms with a unique number of protons in their nuclei.

elementary particle (*or* **fundamental particle**) a particle which is the basic building block for all matter. To explain nuclear interactions fully, the NEUTRON, PROTON and ELECTRON have been supplemented by new particles, some more elementary than others. Two types of particles are thought to exist, namely leptons and hadrons, which are differentiated by the way they interact with other particles. Leptons include the electron and the neutrino, the latter possessing spin but no charge or mass. It is associated with BETA DECAY. Hadrons include protons and neutrons, which are not truly elementary, and it is postulated that these are composed of elementary particles called quarks. Quarks have become part of an elaborate theory of hadron structure encompassing properties termed "flavour" and "colour charge." Although the theory is generally accepted by physicists, quarks have not been confirmed experimentally.

elliptical galaxy a symmetrical GALAXY that does not have spiral arms.

embryo the developmental stage of animals and plants that immediately follows fertilization of the egg cell (ovum) until the young hatches or is born. In animals, the embryo either exists in an egg outside the body of the mother or, as in mammals, is fed and protected within the uterus of the mother.

emersion when a body, for example the MOON, comes out of the shadow that causes its eclipse.

empirical formula a chemical formula of a compound that shows the simplest ratio of atoms present in the compound. For example, the molecule BUTANE (fourth member of the alkane family) has the empirical formula C_2H_5 although its true molecular formula is C_4H_{10}.

emulsion in chemistry, a COLLOIDal solution of one liquid in another.

enantiomorphism the condition when a substance exists in two crystalline forms that are the mirror image of each other (*see* OPTICAL ISOMERISM).

endocrine system the network of glands that secrete signalling substances (HORMONES) directly into the bloodstream. This enables the secreted hormone to travel to, and thus affect, distant target cells, as opposed to just affecting cells that surround the endocrine gland. The pituitary gland (at the back of the head, base of the brain), the thyroid

113

gland (in the neck) and the adrenal glands (above both kidneys) are three of the major endocrine glands within the human body. Each secretes different hormones, which will subsequently affect different parts of the body. For example, the pituitary gland secretes growth hormone, the thyroid gland secretes thyroxine, and the adrenal glands secrete the stress hormone called glucocorticoids, all of which control various functions in the body.

endoplasmic reticulum a network of membrane-bound tubules and flattened sacs connected to the membrane of the nucleus found within all EUCARYOTIC cells. If the endoplasmic reticulum (ER) has RIBOSOMES attached to the outside membrane, then it is called "rough ER," but in the absence of ribosomes it is called "smooth ER," or GOLGI APPARATUS. Both the rough ER and golgi apparatus have functions important in the biosynthesis of other organelles and in the synthesis, modification and sorting of proteins.

endothermic reaction a chemical reaction in which heat energy is absorbed. The required heat energy is supplied by the environment surrounding the reaction, and the products of an endothermic reaction will have stronger bond energies than the original reactants.

energy the capacity to do WORK. There are various forms of energy, including light, heat, sound, mechanical, electrical, kinetic and potential, but all are expressed in the same unit of measurement, called the JOULE (J). Energy has the capacity to change from one form to another (**energy transfer**), but the original input of energy tends to be greater than the final output during energy transfers. As the law of conservation of energy states that it is impossible to make or destroy energy, the difference in the input/output energy levels is a result of the conversion of some of the input energy into an unwanted form, e.g. heat instead of mechanical energy. The energy content of a system or object can be regarded as the "work done" by it and can be calculated using the following equation:

Work Done (w) = Force (F) x Distance Moved (S)

enthalpy the quantity of heat energy (THERMODYNAMICS) possessed by a substance. Enthalpy (H) has units of joules per mole ($Jmol^{-1}$) and is defined by the equation:

$H = E + PV$ where E = internal energy of a system

$\qquad\qquad\qquad\quad P$ = pressure

$\qquad\qquad\qquad\quad V$ = volume

The enthalpy change (δH) during a reaction is referred to as δH negative when the reaction is EXOTHERMIC (heat-evolved) and δH positive when the reaction is ENDOTHERMIC (heat-absorbed).

entropy a measure of the randomness (disorder) of a system. It is a natural tendency of the whole universe that allows all energy to be distributed. Entropy (S) has units of joules per kelvin per mole ($JK^{-1} mol^{-1}$) and can be related to ENTHALPY (H) using GIBB'S LAW ($G = H - TS$). The greater the disorder, the higher the value for entropy, but at absolute zero entropy is also zero.

enzyme any PROTEIN molecule that acts as a natural CATALYST and is found in the bodies of all bacteria, plants and animals. Enzymes are essential for life as they allow the complex chemical reactions of biochemical processes to occur at the relatively low temperature of the body. Enzymes are highly specific in that they will only act on certain SUBSTRATES at a specific pH and temperature. For example, the digestive enzymes called amylase, LIPASE and trypsin will only work in alkaline conditions (pH > 7), whereas the digestive enzyme, pepsin will only work in acidic conditions (pH < 7).

eon the largest geological unit of time (*see* APPENDIX 5), e.g. the Phanerozoic, which includes the Palaeozoic, Mesozoic and Cenozoic ERAS.

ephemeris a published table providing the projected position and movements of planets and comets, of use to a navigator or astronomer.

epicentre the point or line on the Earth's surface that is directly above the focus of an EARTHQUAKE.

epilepsy a seizure disorder caused by lesions in the brain. The symptoms are in the form of attacks, known as fits, which can include a feeling of numbness, muscular convulsions, inability to speak, etc. Epilepsy can be controlled by certain drugs, but in bygone days, surgery was performed on patients who suffered frequent and extreme attacks in an attempt to control these. The operation disconnected the left and right hemispheres of the brain by cutting the communication system between them, a fibrous network called the *corpus callosum*.

epoch an interval of geological time (*see* APPENDIX 5) subsidiary to the period, and forming several ages, e.g. the Pleistocene.

equatorial an astronomical telescope that is positioned so that when it is set on a star, that body will be kept in the field of view. This is achieved by mounting it to revolve about an axis parallel to the Earth's axis.

equilibrium a condition in which the proportion of

reactants and products is constant as the rate of the forward reaction equals the rate of the reverse reaction. An equilibrium constant (K_c) can be calculated for reactions by dividing the concentrations of products from the forward reaction by the concentration of the reactants from the reverse reaction. Both the concentration of the products and reactants are raised to the power corresponding to their coefficient in the chemical equation for the reaction, i.e. for the reaction:

$$aA + bB \rightleftharpoons xX + yY$$

the equilibrium constant (K_c) is calculated as follows:

$$K_c = \frac{[X]^x \, [Y]^y}{[A]^a \, [B]^b}$$

K is expressed in moles per litre (mol^{-1}) and is useful for indicating whether a particular reaction is irreversible (large K_c) or the reactants are unreactive (small K_c). Equilibrium is affected by changes in pressure, temperature and concentration, but is not affected by the addition of a CATALYST (this affects only the reaction rate).

equinox the point of intersection between the SUN's *apparent* path in the sky (relative to the stars) and the CELESTIAL EQUATOR. As the Sun physically crosses the celestial equator north to south, it is the autumnal equinox, and south to north is the vernal equinox.

equivalent weight *or* **chemical equivalent** the quantity of a substance that reacts with a certain amount of a standard. For an ELEMENT, the equivalent is the number of grams that combine with or displace 8 grams of oxygen (or 1 gram of hydrogen).

The equivalent weight of a base is the amount required to neutralize the equivalent weight of an acid.

ER *see* **endoplasmic reticulum**.

era a unit of geological time that comprises several PERIODS, for example, the Palaeozoic (divided into upper and lower) contains six periods, from the Cambrian to the Permian.

erosion the destructive processes that, with WEATH-ERING, constitute DENUDATION. Specifically, erosion involves the further breakdown and transport of material by water, ice and wind, and in the transportation the continued wearing down of land surfaces. The transporting agents thus erode by wind laden with sand scouring rock, glaciers containing rocks and boulders grinding down the rocks over which they pass, and rivers excavating their own courses due to movement of rocks, pebbles and particles in the water. Rivers also carry materials in solution.

erratic a rock that is moved by a glacier and deposited on a landscape of totally different rock, showing that the former could not have been derived from the latter.

erythrocyte the red blood cell of vertebrates that is made in the bone marrow. It differs from other cells in the human body in that just before it is released into the bloodstream, it sheds its nucleus. Erythrocytes contain HAEMOGLOBIN, the protein molecule essential for transportation of oxygen from the lungs to all tissues in the body. Within the erythrocyte membrane there is a complex molecule called the Band 3 protein, which is essential for the transport of carbon dioxide (CO_2) from all body tissue to the lungs as it allows CO_2, in the form of the bicarbonate ion (HCO_3^-), to leave the red blood cell in exchange for a chloride ion (Cl^-). A person deficient in the number of erythrocytes circulating in their blood is said to be anaemic. Anaemia can be a result of a defect in the structure of the erythrocyte membrane or lack of iron to form the haem group of the haemoglobin molecule.

escape velocity the velocity required of an object, e.g. a rocket, space probe, etc, to escape from the gravitational pull of a larger body, e.g. a planet. The necessary velocity depends upon the planet's mass and diameter. For the EARTH it is 11,200 ms^{-1} (or 25,000 miles per hour) and for the Moon, 2400 ms^{-1}. The behaviour of a BLACK HOLE is explained by its escape velocity being greater than the SPEED OF LIGHT.

escarpment *or* **scarp** a cliff or steep slope at the edge of an essentially flat or gently sloping area, generated by a combination of original geology and the attitude of the rocks, and subsequent erosion.

esker a steep-sided ridge composed of sands and gravels, the former showing CROSS-BEDDING. It is the remains of a stream that ran beneath or within a glacier.

ESR *see* **electron spin resonance**.

ester an organic compound formed from acids by replacing the hydrogen with an alkyl radical, e.g. CH_3COOH (ethanoic acid) to $CH_3COOC_2H_5$. Many esters have a fruity smell and are used for flavourings. Esters are common in nature, as animal fats and vegetable oils are formed from mixtures of esters.

estuary a partially enclosed stretch of water that is

subjected to marine tides and fresh water draining from land. An estuary is usually created as a DROWNED VALLEY, due to a post-glacial rise in sea level. A large amount of sediment is deposited in estuaries, and the tidal currents may produce channels, sand-banks and sand waves.

ethane the second member of the homologous series of ALKANES. It is an insoluble colourless gas with chemical formula CH_3CH_3. Ethane has no reactive functional group, and is, therefore, a stable molecule that will not react with ACIDS, BASES, ELECTROPHILES or NUCLEOPHILES, but, however, it will undergo a slow SUBSTITUTION reaction.

ethanoic acid *see* **acetic acid**.

ethanol a derivative of ETHANE, which has a functional hydroxyl group (OH) in place of one hydrogen atom, i.e. CH_3CH_2OH. Ethanol is obtained either by fermentation of carbohydrates to form alcoholic beverages, or commercially prepared from ETHENE by adding water and sulphuric acid.

ethene the first member of the ALKENE family, which is an insoluble gas at room temperature. Ethene (C_2H_4) is an important precursor in the industrial manufacture of the plastic POLYMER called poly-thene, i.e. polyethene.

ethylene glycol (dihydroxyethane $HOCH_2CH_2OH$) a colourless, HYGROSCOPIC liquid used in anti-freeze liquids and as a coolant for engines. It is used for the manufacture of polyester fibres, such as Terylene, and elsewhere in the textile industry, in printing inks, and foodstuffs.

ethyne the first member of the ALKYNE family, it has the chemical formula C_2H_2. It is also known as acetylene and is a highly flammable gas which, when burned with oxygen, will produce the high temperature flame (> 2500°C) characteristic of the oxyacetylene torch needed to cut and weld metals.

eucaryote any member of a class of living organisms (except viruses) that has a membrane-bound nucleus within its cells. All eucaryotic cells contain OR-GANELLES, which are also bound by closed, phospholipid membranes, e.g. chloroplast, endoplasmic reticulum, mitochondrion etc. All plants and animals are eucaryotes, but bacteria and cyanobacteria are PROCARYOTES.

eugenics the study of how the inherited characteristics of a human population can be improved by genetics, i.e. controlled breeding.

eustatic movements world wide changes in sea level. These may be due to changes in glaciers, or tectonics, e.g. large crustal movements in ocean basins.

eutectic pertaining to a mixture (SOLID SOLUTION) of two or more substances that has a melting point lower than that of either component. This is a eutectic mixture and applies in particular to mixtures of metals but also to minerals.

evaporation the process by which a substance changes from a liquid to a vapour. Evaporation occurs when a liquid is heated and some molecules near the surface of it eventually have enough KI-NETIC ENERGY to overcome the attractive forces of the remaining molecules and escape into the surrounding atmosphere. During evaporation from an open container, the temperature of the liquid falls until heat from the surroundings flows in to replace this heat loss. This explains why swimmers can feel chilled when they emerge from the water; heat energy from the skin is converted to kinetic energy, allowing some water molecules to evaporate.

evaporite sedimentary rock formed by precipitation from solution during evaporation of lagoons, salt pans (a basin area in a semi-arid region), and saline lakes. The least soluble salts, such as CALCIUM CARBONATE and magnesium carbonate, precipitate first, followed, for example, by (in increasing solubility) sodium sulphate, and then potassium chloride and magnesium sulphate. The same principle applies to evaporating sea water, when calcium sulphate (or GYPSUM) is the first to precipitate, followed by ANHYDRITE and HALITE (NaCl). The rock types formed, in addition to gypsum and anhydrite, include LIMESTONE, DOLOMITE and rock salt (halite).

evening star in general terminology, the name given to Venus (or Mercury), which is seen in the western sky around sunset.

event horizon the boundary of a BLACK HOLE, which is considered to be a sphere. This marks the line beyond which light cannot escape.

evolution the process by which an organism changes and thus attains characteristics distinct from existent relatives. Any species of organism will only evolve if:

(a) There has been genetic mutation allowing variation in the genetic information the parent passes on to its descendants.

(b) An individual proves to be more suitable to a particular environment than its relatives, allowing it to survive and propagate whereas its relatives will become extinct, i.e. NATURAL SELECTION (*see also* DARWINISM).

exfoliation the process whereby rocks are gradually

worn away by the flaking off of layers or shells. The process involves a number of factors, including chemical WEATHERING, which breaks down and may cause expansion of some minerals, and repeated expansion or contraction due to temperature changes.

exothermic reaction a chemical reaction in which heat energy is released to the surrounding environment. The products of an exothermic reaction will have weaker bond energies and therefore be more stable than the bonds within the molecules of the original reactants.

exponent a symbol, usually numerical, that appears as a superscript to the right of a mathematical expression and indicates the power to which the expression has to be raised. For example, the expressions a^5 and 3^7 have the exponents 5 and 7 respectively.

exponential function a mathematical function in which the constant quantity of the expression is raised to the power of a variable quantity, i.e. the exponent. For example, the exponential function, $f(x) = a^{2x}$, has 2x as its variable exponent. However, the term exponential function mainly refers to the function $f(x) = e^x$, in which e has a value equal to 2.7182818. The exponential function e^x is the inverse of the natural LOGARITHMIC FUNCTION, ln(x).

exponential series the sum of the infinite sequence of exponential terms for either the real exponential function (e^x) or the complex exponential function (e^z). It is defined by the following:

$$e^x = \sum_{r=0}^{\infty} \frac{x^r}{r!} = 1 + \frac{x}{1!} + \frac{x^2}{2!} + \frac{x^3}{3!} + \cdots + \frac{x^r}{r!} + \cdots$$

In the above, X represents a real function and

extrapolate

therefore the function e^x will tend to 0 as X tends to negative infinity ($-\infty$), that is, as:

$$X \longrightarrow -\infty, e^x \longrightarrow 0.$$

The term complex function denotes the presence of a complex number ($z = a + ib$) within the function.

extrapolate to predict the unknown value of a measurement or function using known values. On a graph, extrapolation involves extending the curve of the function beyond the set of known values for the x and y co-ordinates.

extrusive rocks a general term to encompass rocks of volcanic origin that are discharged onto the Earth's surface.

F

fabric the physical/geometrical arrangement of minerals or structural elements in a rock. In general, fabric refers to the microscopic or medium scale, and may be planar or linear.

facies a term used of IGNEOUS, SEDIMENTARY and META-MORPHIC rocks to encompass all the features that characterize a rock type and thus the conditions under which it was formed. The features may relate to mineral composition, fauna or sedimentary structures in a sedimentary facies. With METAMORPHIC rocks, the concept of facies is slightly different and refers to sets of mineral assemblages formed from different rock types that have been subjected to the same metamorphic grade (conditions).

factor one of two or more quantities that produce a given quantity when multiplied together. For example, the factors of the number 8 are 1,2,4,8.

factorial the product of a series of consecutive INTE GERS from 1 to n inclusive, where n is a whole number. Thus, factorial 5, written as 5! = 1 x 2 x 3 x 4 x 5 = 120. Factorials of much larger numbers are not usually defined.

faculae (*singular* facula) large areas within the SUN's surface that are brighter than their surroundings and exhibit much higher temperatures.

Fahrenheit (F) a temperature scale, devised by the

125

German physicist Gabriel Fahrenheit (1686-1736), that sets the freezing point of water at 32° and the boiling point at 212°. Fahrenheit temperatures can be converted to CELSIUS by the equation

$$F = 1.8C + 32.$$

fall-out radioactive material deposited on the ground from the atmosphere. The source of the radioactive substances is nuclear explosions or escape from nuclear reactors, whether through failure to adequately filter coolants or because of an accident.

fan an area of detrital sediment occurring in a submarine environment at the base of cliffs or mountains. An alluvial fan is a sediment load deposited when a stream slows (and therefore loses its carrying power) caused by a decrease in gradient, e.g. on reaching a plain having flowed through a mountain range. Also CLEAVAGE fan, where cleavage planes form fans in folded rocks.

farad the unit of CAPACITANCE, usually denoted by the symbol F. For example, a capacitance (C) such that a charge of one COULOMB raises the potential to one volt, is said to be one farad, i.e. 1 farad = 1 coulomb per volt ($1F = 1\ CV^{-1}$). As the farad itself is usually too large a quantity for most applications, the practical unit is the microfarad (μF).

faraday the quantity of charge carried by one mole of electrons ($\approx$ Avogadro's constant x charge on an electron), with the value 9.6487×10^4 coulombs. It was named after Michael Faraday (1791–1867), a British scientist whose contributions to physics and chemistry include ELECTROMAGNETIC INDUCTION, electrolysis and MAGNETIC FIELDS.

fatigue of metals the structural failure of metals due to the repeated application of STRESS, which results in a change to the crystalline nature of the metal.

fats a group of naturally existing lipids that occur widely in plants and animals and serve as long-term energy stores. A fat consists of a GLYCEROL molecule and three FATTY ACID molecules, collectively known as a triglyceride, which is formed during a condensation reaction (water is released). Fats are important as energy-storing molecules since they have twice the calorific value of carbohydrates. In addition, they insulate the body against heat loss and provide it with cushioning, which helps protect against damage. In mammals, a layer of fat is deposited beneath the skin (subcutaneous fat) and deep within the tissues (adipose tissue) and is solid at body temperature due to the high degree of saturation. In plants and fish, the fatty acids are generally less saturated and as such tend to have a liquid-like consistency, i.e. oils, at room temperature.

fatty acids a class of organic compounds containing a long hydrophobic (water insoluble) hydrocarbon chain and a terminal carboxylic acid group (COOH) which is extremely hydrophilic (water soluble). The chain length ranges from one carbon atom (HCOOH; methanoic acid), to nearly thirty carbon atoms, and the chains may be SATURATED or UNSATURATED. As chain length increases, melting points are raised and water solubility decreases. However, both unsaturation and chain branching tend to lower melting points. Fatty acids have three major physiological roles:

(1) They are building blocks of phospholipids (lipids containing phosphate) and glycolipids (lipids con-

taining carbohydrate). These molecules are important components of biological membranes, creating a lipid bilayer which is the structural basis of all cell membranes.

(2) Fatty acid derivatives serve as hormones and intracellular messengers.

(3) Fatty acids serve as fuel molecules. They are stored in the CYTOPLASM of many cells in the form of triglycerides (three fatty acid molecules joined to a glycerol molecule) and are degraded, as required, in various energy-yielding reactions.

fault a planar fracture caused by brittle deformation across which rocks are displaced, with the displacement usually parallel to the fracture plane. Movement along faults may vary from millimetres to kilometres. The measurement of the movement is often made by referring to the horizontal and vertical components (*see* HEAVE and THROW; *also* HADE). There are several types of fault, depending upon the sense of movement, including STRIKE-slip and dip-slip faults and THRUSTS.

fault breccia a zone composed of broken, angular and crushed rock fragments generated by movement along a fault. A loose-textured (or incohesive) rock, at depth a fault BRECCIA may become lithified with some recrystallization to form a crush breccia.

feedback mechanism a control mechanism that uses the products of a process to regulate that process by activating or repressing it. Almost all homeostatic mechanisms (*see* HOMEOSTASIS) in animals operate by negative feedback, whereby a variation from the normal triggers a response that tends to oppose it. For example, it operates during hormonal release to maintain steady blood sugar levels. Positive feedback is found less often as a biological control mechanism. Here, a variation from the normal causes that variation to be amplified, and this is usually a sign that the normal control mechanisms have broken down.

feldspar a very important group of rock-forming minerals that are aluminium silicates with combinations of calcium and sodium (Ca/Na) or sodium and potassium (Na/K). The variation in chemical composition produces two series that are SOLID SOLUTIONS. The alkali feldspar, containing end members with Na or K, are common in igneous rocks, and the potassic variety, orthoclase, forms characteristically pink crystals in some GRANITES. The second solid solution forms the plagioclase feldspars, with end members albite (sodic) and anorthite (calcic), abbreviated Ab and An. Plagioclase feldspar occurs across a wide range of rock types and is useful in their classification. Intermediate members between Ab and An are classified by the percentage of the calcic end member present. The different plagioclase feldspars are identified by study of THIN SECTIONS under the microscope.

feldspathoids a group of minerals that, as the name suggests, resemble FELDSPARS chemically and structurally but contain less SILICA. Minerals in this group crystallize from MAGMAS deficient in silica and therefore never occur with quartz (which is "free" silica), and are thus found in rocks low in silica and rich in alkalis.

fermentation a form of ANAEROBIC RESPIRATION, which converts organic substances into simpler molecules, generating energy in the process. Fermentation, carried out by certain organisms such as bacteria and YEASTS, is the conversion of sugars to alcohol in the process known as alcoholic fermentation. Lactic acid fermentation occurs in the muscles of higher animals when the oxygen requirement exceeds the supply and sugar is converted into lactic acid. In industry, fermentation is important in baking and in beer and wine production, and these use large quantities of yeast.

Fermi, Enrico (1901–1954) an Italian-born American physicist who was awarded the Nobel Prize in 1934 for his discovery that stable elements would become unstable when bombarded with NEUTRONS as they have become radioactive. His later research contributed to the harnessing of atomic energy and to the construction of the atomic bomb.

ferromagnetism an enhanced magnetism found in some materials that are highly susceptible to magnetization. Iron, nickel, cobalt and some alloys, e.g. iron-nickel, are more magnetic than other substances because of this phenomenon. Essentially it is due to the electronic structure of the metal elements, which have unbalanced electron spin in their inner electron orbits, producing a certain magnetization. The application of an external field produces magnetism that varies with the applied field.

ferruginous deposits SEDIMENTARY rocks that contain iron in sufficient quantities to be exploited as iron ore. The iron may occur in different mineral forms: the carbonate siderite, which often alters to oxide ores; the silicate chamosite; the oxides magnetite and haematite; or the hydrated oxide LIMONITE (essentially a rock composed of the mineral goethite). These minerals may form in numerous ways, from

filtration

precipitation, to weathering or alteration products, or primary segregations. They also occur as components in IGNEOUS and METAMORPHIC rocks.

fertilization the fusion of male and female GAMETES to produce a single cell, which sets in motion a chain of events that gives rise to a new individual. In animals, where the gametes unite outside the parents' bodies, it is termed external fertilization (as in most fish). Where the male gametes are deposited within the body of the female by the male, it is termed internal fertilization, as is the case with mammals. In flowering plants, after pollen has been transferred from the male to the female part of the flower, a pollen tube develops, which transfers two male nuclei to the ovule of the female. Double fertilization occurs, producing a DIPLOID ZYGOTE and a triploid endosperm, which act as a food supply for the developing embryo.

fertilizer a substance added to soil to replace nutrients removed by plants, thus contributing towards their health and vitality. Fertilizers may be natural (e.g. manure) or synthetic, the latter containing nitrogen, phosphorus and potassium as the main constituents.

fibrinogen a blood PROTEIN, which causes BLOOD CLOTTING due to action by the ENZYME thrombin. The end product is fibrin.

field *see* **electric field, magnetic field**.

filter paper a pure cellulose paper used in the laboratory for the separation of solids from liquids by FILTRATION.

filtrate the liquid remaining after FILTRATION, having been separated from a solid or liquid mixture.

filtration the separation of a solid from a liquid by

131

passing the mixture through a suitable separation medium, e.g. FILTER PAPER, that holds back the solid and permits the liquid to pass through.

finder a small telescope used in conjunction with, and fixed to, a large telescope, for finding the required object and fixing it in the centre of the viewing field of the large telescope.

fining-upwards cycle (*see* TURBIDITE) a cyclical sedimentation where coarse-grained material grades upwards, gradually, into material of finer grain size.

fireclay a clay rich in SILICA and alumina (Al_2O_3) and low in iron, magnesium and calcium, which, because of its high melting point, > 1600°C, is used widely for refractory purposes, e.g. crucibles, furnaces. They occur commonly beneath coal seams, as SEAT EARTHS.

firn *see* **névé**.

fissile a fissile element is one that undergoes NUCLEAR FISSION.

fission the spontaneous or induced splitting of a heavy nucleus (such as URANIUM) into two fragments during a nuclear reaction, which subsequently releases vast quantities of energy. Nuclear fission is induced by irradiating nuclear fuels like uranium with NEUTRONS in a device called a nuclear reactor, and this process is accompanied by the emission of several neutrons. These neutrons in turn cause fission of another nucleus, which, under suitable conditions, can result in a CHAIN REACTION. The energy released is in the form of heat, and it can be harnessed and used to produce electricity by making steam, which is used to drive turbines.

fission track dating a technique used to date some

minerals and natural glasses. The FISSION of ^{238}U creates charged particles, which leave a trail through the solid glass. The tracks can be studied using a microscope, and their number relates to the age of the specimen and its uranium concentration, providing certain physical criteria have been met.

flagella (*singular* **flagellum**) long thread-like extensions from the surface of a cell, similar to cilia. Protozoa (single-celled animals) use flagella for locomotion.

flame test a simple test for the detection of metals and useful for distinguishing between different metals. A small quantity of the unknown sample is placed on a platinum wire and held in a flame, and the resultant colour is characteristic of the particular metal. For example, when sodium compounds are held into a flame, the flame burns with a bright yellow colour. Potassium gives a violet flame, and lithium and strontium give a red flame. Although lithium and strontium appear similar, the light from each can be resolved (separated) into different colours by using a prism, and this resolution easily distinguishes the two elements.

flare a sudden outburst of radiation from a star, or more particularly from the lower atmosphere of the SUN.

flaser structure a rock produced by intense pressure, where large crystals in rock lenses occur in a fine-grained, streaky GROUNDMASS.

flash distillation a technique that involves spraying a liquid mixture into a heated chamber at low pressure. This enables rapid removal of solvent due to its greater volatility. The technique is used in the petroleum industry.

Fleming, Sir Alexander (1881–1955) a Scottish bacteriologist who discovered the antibiotic penicillin in 1928, for which he was awarded the Nobel Prize in 1945.

flint a form of SILICA, composed of minute crystals (*see also* CHALCEDONY), occurring commonly as nodules or bands in CHALK deposits. It breaks with a conchoidal (curved) fracture to produce sharp edges, a property utilized by prehistoric man for the fabrication of tools and weapons.

flood basalt basaltic lavas that erupt from a fissure in the continental crust under tension and cover vast areas of country, a veritable "flood" of lava. There are numerous examples of flood basalts, including East Greenland, the Deccan region of India, and large areas of southern Africa. It seems that these areas, which may reach up to half a million square kilometres (nearly 200,000 square miles) or more and thicknesses of several kilometres, were emplaced in (geologically) a relatively short time span (a few million years).

flood plain the planar or near-planar surface at the bottom of a river valley within which the river flows and which is covered when the river floods. It is formed by the progressive development of the river as it MEANDERS laterally. It is made up of river-borne sediment, which often shows a fining-upwards sequence from bedrock, through coarse gravels, then sand, followed by clays and silts. This rhythmic sequence of sediments may be repeated.

flow structure a feature often seen in viscous MAGMAS, with contorted banding of differing colour or alignment of crystals.

fluid any substance that flows easily and alters its

shape in response to outside forces. All gases and liquids are fluids. In liquids, the particles move freely but are restricted to the one mass, which occupies almost the same volume. In gases, however, the particles tend to expand to the limits of their containing space and thus do not keep the same volume.

fluid-mosaic the name given to the model that describes the structure of the cell membrane in organisms, proposed by Singer and Nicholson in the 1970s. Using electron microscopy, they confirmed that the lipid component is organized in a regular bimolecular structure with protein molecules arranged irregularly along the lipid layers. Both lipid components can move laterally in the membrane.

fluorescence the property of certain substances to re-emit absorbed radiation as visible light. This occurs when molecules in the ground state are excited by incident light of a particular WAVELENGTH, thus raising their energy level. When the energy level falls, the radiation is emitted at a different, usually greater, wavelength.

fluorocarbons (*see also* CFC) a group of compounds where fluorine replaces some or all of the hydrogen atoms in a hydrocarbon. The resulting compounds are similar to the original hydrocarbon in some ways, but are unreactive and thermally stable. Uses include oils and greases, fluorine-containing polymers, SURFACTANTS, and also refrigerants and aerosol propellants, although these latter uses are being discontinued because of the effect of CFCs on the OZONE LAYER.

fluviatile deposits sediments deposited by a river, often comprising sands and gravels.

flux a substance added to a solid to assist in its fusion, e.g. cryolite (naturally occurring sodium aluminium fluoride) is added to bauxite from which ALUMINIUM is extracted.

flysch a flysch sequence comprises many graded beds (*see* GRADED BEDDING) alternating with shales, the former occurring as a TURBIDITY CURRENT. The term flysch originated from the Alps, where rock sliding occurs.

focal plane in an optical system, the plane in which the image of an object is formed and therefore the position for the film or plate. The focal plane is at right angles to the principal axis of a lens (system), which runs through the centre of the lens, and at 90° to the long dimension of the lens.

focal point the point at which converging rays meet on the axis of a lens system.

focus the point at which rays converge after passing through an optical system.

foetus the developing young in the mammalian uterus, from the post-embryonic period until birth. In humans, this period is from about 7-8 weeks after FERTILIZATION.

fog a suspension of water droplets (up to 20µm diameter) in the atmosphere when the air is near to saturation, causing visibility to fall below 1 km (3280 ft). Fog formation is enhanced by the presence of smoke particles, which act as nuclei for the condensation of the droplets. The condensation may be caused by cooling of the ground or warm air moving over cold ground or water.

föhn wind a general term for a warm, dry wind in the lee of a mountain range.

fold a fold is produced when originally planar rock

surfaces become curved or bent due to ductile ("flowing") deformation. Study of fold styles in areas of tectonism enables much to be understood about the mechanisms and direction of the forces involved.

There are many types of fold, depending on diverse factors such as rock type, intensity of deformation, etc, but there is a basic descriptive morphology that can be applied. The zone of maximum curvature is the hinge, and the limbs lie between hinges or (for a single fold) on either side of the hinge. Thus, in a series of folds each limb is common to two adjacent folds. Two other useful parameters are the axial plane and axis. The axial plane for one fold is an imaginary feature that bisects the angle between both limbs and is equidistant from each limb. The fold axis is the line created by the axial plane intersecting the hinge zone. Folds range in size from a few millimetres to many hundreds of metres.

foliation (*see also* CLEAVAGE *and* SCHIST) a planar fabric created in a rock due to deformation. It is a general term covering several such fabrics, including cleavage, fracture cleavage (created by closely spaced fractures), and schistosity. Folds and foliation formed by the same period of deformation will exhibit a geometrical relationship to each other, and often the foliation is parallel or sub-parallel to the axial plane of the trends of major folds.

folic acid a compound that forms part of the VITAMIN B complex. It is involved in the BIOSYNTHESIS of some AMINO ACIDS and is used in the treatment of anaemia.

footwall in faulting, and where the FAULT plane is inclined, the "block" below the fault is referred to as the footwall block. (The "block" above the fault is called the hanging-wall block.)

force the push or pull exerted on a body, which may alter the state of motion by causing the velocity of the body to increase or decrease. An object will continue to move at a constant speed in a straight line unless another force acts upon it. The unit of force is the newton, given by $F = ma$ where m is the MASS of the body and a is its ACCELERATION.

formation the basic rock unit used in lithostratigraphy (description of rocks in terms of their LITHOLOGY). Formations are divided into members, and several formations form a group.

formula (*plural* **formulae**) a law or fact used in science and mathematics, denoted by certain symbols or figures. In mathematics and physics, it is expressed in algebraic or symbolic form. In chemistry, there are three principal types of formulae— EMPIRICAL FORMULA, MOLECULAR FORMULA, and STRUCTURAL FORMULA.

fossil the remains of once-living plants and animals, or evidence of their existence, preserved in the strata of the earth's crust. Palaeontology is the name given to the study of fossils and has proved useful in the study of evolutionary relationships between organisms, and in the dating of geological strata.

fossil fuels these are NATURAL GAS, PETROLEUM (oil) and coal, which are the major fuel sources today. They are formed from the bodies of aquatic organisms that were buried and compressed on the bottoms of seas and swamps millions of years ago. Over time, bacterial decay and pressure converted this organic matter into fuel.

Hard coal, which is estimated to contain over 80 per cent carbon, is the oldest variety and was laid down up to 250 million years ago. Another, younger variety (bituminous coal) is estimated to contain between 45 per cent and 65 per cent carbon. The fuel values of coal are rated according to the energy liberated on combustion. Coal deposits occur in all the world's major continents, and some of the leading producer countries are the United States, China, Russia, Poland and the United Kingdom.

Natural gas consists of a mixture of HYDROCARBONS, including METHANE (85 per cent), ETHANE (about 10 per cent) and PROPANE (about 3 per cent). However, other compounds and elements may also be present, such as carbon dioxide, hydrogen sulphide, nitrogen and oxygen. Very often, natural gas is found in association with petroleum deposits. Natural gas occurs on every continent, the major reserves being found in Russia, the United States, Algeria, Canada and in counties of the Middle East.

Petroleum is an oil consisting of a mixture of HYDROCARBONS and some other elements (e.g. sulphur and nitrogen). It is called crude oil before it is refined. This is done by a process called FRACTIONAL DISTILLATION, which produces four major fractions:

(1) Refinery gas, which is used both as a fuel and for making other chemicals.

(2) Gasoline, which is used for motor fuels and for making chemicals.

(3) Kerosine (paraffin oil), which is used for jet aircraft, for domestic heating and can be further refined to produce motor fuels.

(4) Diesel oil (gas oil), which is used to fuel diesel engines.

The known residues of petroleum of commercial importance are found in Saudi Arabia, Russia, China,

Kuwait, Iran, Iraq, Mexico, the United States, and a few other countries.

Together, the fossil fuels account for nearly 90 per cent of the energy consumed in the United States. As coal supplies are present in abundance compared with natural gas or petroleum, much research has gone into developing commercial methods for the production of liquid and gaseous fuels from coal.

fossilization the formation of a FOSSIL. Organisms tend to undergo some changes after death and are not usually preserved whole. In particular, the soft parts will decay and skeletal parts are often changed. Sediment may flatten an organism, and porous structures may be replaced by minerals. Recrystallization commonly occurs, replacing the fine structure of shells, and the mineral aragonite (a form of CALCIUM CARBONATE) often changes to the more stable CALCITE. Skeletal parts may leave impressions or moulds in sediments, and these may be internal or external. In addition, burrows, trails and similar evidence of organisms can be preserved as TRACE FOSSILS.

fraction a quantity that is only part of a whole unit. It is written as $^x/_y$ where x is an integer and y is a natural number.

fractional crystallization *see* DIFFERENTIATION.

fractional distillation (*also called* **fractionation**) the process used for separating a mixture of liquids into component parts (fractions) by distillation. The liquid to be separated is placed in a flask or distillation vessel to which a long vertical column (fractionating column) is attached. The liquid is boiled, causing it to vaporize, and as the vapour rises up the column it condenses and runs back into the vessel. The vapour in the vessel continues to rise, and as it does so it passes over the descending liquid. This eventually creates a steady temperature gradient, with temperature decreasing towards the top of the column. The components of the mixture that vaporize easily (low BOILING POINTS) are said to be more volatile and are found towards the top of the column, where the temperature is lowest, while less volatile components are found towards the bottom of the column. At various points on the column, the different fractions can be drawn off and collected. Those components with appreciably different boiling points will be separated into the different fractions. Petroleum contains a mixture of hydrocarbons, and fractional distillation is used to separate the components into fractions such as gasoline and kerosine. It was fractional distillation of liquid air that led to the discovery of three of the NOBLE GASES (neon, krypton and xenon), and today the process is used to obtain large quantities of molecular oxygen required for commercial purposes. Air is first compressed and cooled (which freezes out carbon dioxide and water) and then fractioned in a liquid air machine. Molecular oxygen can be obtained since the other components of air (nitrogen and argon) are more volatile and can be removed from the top of the fractionation column as gases.

fracture a general term applied to a break in any material. Specifically in mineralogy, it refers to breaks in minerals (and rock) that are not due to CLEAVAGE or FOLIATION. Most minerals have an UNEVEN fracture, but some display a characteristic *conchoidal* fracture (as in flint) of curved convex or concave surfaces. These often occur as a concentric feature diminishing to the point of impact (similar in fashion to shell growth lines).

fraternal twins unidentical twins (dizygotic twins) that develop when two ova are fertilized simultaneously. This occurs when two ova have matured and have been shed simultaneously, and the resultant twins resemble each other only to the same extent as brothers and sisters born at different times.

Fraunhofer lines in the continuous spectrum of the Sun, Fraunhofer lines occur as dark lines caused by the absorption of certain wavelengths of light by elements in the CHROMOSPHERE. The major lines are due to the presence of calcium, hydrogen, sodium and magnesium.

free energy (*also called* **Gibbs free energy**) a thermodynamic quantity used in chemistry, which gives a direct criterion of spontaneity of reaction in a reversible process. It is defined by the equation

$$G = H - TS,$$

where G is the energy liberated or absorbed, H is the ENTHALPY, S is the ENTROPY, and the system is measured at constant pressure and temperature (T).

As a reaction proceeds, reactants form products, and H and S change. These changes, denoted by DH and DS, result in a change in free energy, DG, given by the equation DG = DH – TDS. If DG is a large negative number, the reaction is spontaneous, and reactants transform almost entirely to products when equilibrium is reached. If DG is a large positive number, the reaction is not spontaneous, and reactants do not give significant amounts of products at equilibrium. If DG has a small negative or positive value (less than 10 kilojoules), the reaction gives a mixture of both reactants and products in significant amounts at equilibrium.

freeze-drying a process used when dehydrating heat-sensitive substances (such as food and blood plasma) so that they may be preserved without being damaged during the process. The material to be preserved is frozen and placed in a VACUUM. This causes a reduction in pressure, which in turn causes the ice trapped in the material to vaporize, and the water vapour can be removed, producing a dry product. For most solids, the pressure required for vaporization is quite low. However, ice has an appreciable vapour pressure, which is why snow will disappear in winter even though the temperature is too low for it to melt.

frequency (f) the number of complete wavelengths passing any given reference point on the line of zero disturbance in one second. For example, the frequency of an OSCILLATION, such as a wave, is the number of complete cycles produced in one second, the unit of which is the hertz (Hz). The wave equation is given by $c=f\lambda$, where c = the velocity of the wave and λ is its wavelength. For example, if waves have a wavelength of 2 metres and travel with a velocity of 10 metres per second, then the frequency of the wave motion is 5 Hz.

friction the force that opposes motion and always acts parallel to the surface across which the motion is taking place. Unless a force is exerted to keep an object moving, it will tend to slow down due to the opposing force of friction. Friction can therefore be thought of as a negative force, causing negative acceleration. Frictional forces between two solids or between a solid and a liquid are much greater than those between a solid and air. The hovercraft is a vehicle that exploits this fact by travelling on a cushion of air, thus reducing friction.

friction layer the layer in the atmosphere in which the effects of surface friction are registered. It extends to approximately 610 metres (2000 feet).

front the boundary between large masses of air with different properties. Fronts can be identified by the air masses separated at the front, their stage of development, and the direction of their advance (*see* COLD FRONT *and* WARM FRONT).

fuel any material that, when treated in a particular way, releases energy in the form of heat. The FOSSIL FUELS and organic fuels (e.g. wood and waste material) produce energy by combustion in the presence of oxygen, releasing carbon dioxide and water. Nuclear fuels, such as uranium and plutonium, release large amounts of heat during nuclear reactions when chemical changes occur within the atom (*see* FISSION).

fugacity a quantity, used in chemical equilibrium calculations, that relates to gas being likely to expand or escape.

fuller's earth a clay consisting primarily of montmorillonite (a complex silicate of aluminium with water in its structure) used originally to absorb fats from wool (termed "fulling," hence the name). Fuller's earth is now used in the textile industry and in the refining of oils and fats.

fumarole a vent in a volcanically active area from which is emitted high-temperature gases (often H_2O, CO_2 and SO_2 in the main). It may indicate a late stage in volcanic activity although it can precede eruptions.

function a mathematical term used when there is a relationship between two or more VARIABLES. For example, if two variables are x and y, and there is an associated value for y for every value of x, then y is said to be a function of x. The values of x are termed the domain of such a function, and the range of that function is the term given to the corresponding set of y values.

functional group an arrangement of atoms joined to a carbon skeleton, which gives an organic compound its chemical properties. Compounds with the same functional groups are classed together because of their similar properties. For example, compounds with the functional group NH_2 are classed as amides; those with carbon-carbon double bonds are classed as ALKENES; those with the -OH group are classed as ALCOHOLS.

fundamental particle *see* **elementary particle**.

fungus (*plural* **fungi**) simple unicellular or filamentous plants with no chlorophyll. Fungi cause decay in fabrics, timber and food, and diseases in some plants and animals. Particular fungi are used in brewing, baking, and in the production of ANTIBIOTICS.

furan resins plastics formed from furan (C_4H_4O) and used widely as plastic coatings on metal and for adhesives.

fuse a device for maintaining the CURRENT in a CIRCUIT by preventing it from rising too high if a fault should occur. It is simply a thin metal wire of low MELTING POINT so that the heat generated from too high a current melts the wire, causing the circuit to break and the current to fall to zero. Fuses have different ratings according to the thickness of the wire. The fuse rating is the maximum current that can flow through the fuse without causing the circuit to break.

fusion a nuclear reaction in which unstable nuclei combine to create larger, more stable nuclei with the release of vast amounts of energy. For the reaction to occur, the nuclei have to collide, and this requires the nuclei to have very high KINETIC energies to overcome the repulsive forces between them. NUCLEAR FUSION occurs in the hydrogen bomb (fusion bomb), and at temperatures of about 100 million degrees centigrade, the reaction becomes self-sustaining.

G

galactic circle the great circle created by the intersection of the GALACTIC PLANE and the CELESTIAL SPHERE. (The great circle is simply the cutting of a sphere by a plane that passes through its centre.)

galactic halo a collection of stars, gas and dust, nearly spherical in shape and having a common centre with the GALAXY. It contains population II stars (*see* POPULATION TYPES) and creates a lot of the background radio emissions.

galactic noise background noise from galactic sources.

galactic plane the plane that passes through the centre of the Milky Way, as far as that can be determined.

galactic rotation the rotation of the GALAXY, and all its gas, stars and dust, which increases towards its centre. The rotation near the SUN is approximately $250 \, kms^{-1}$ (155 miles), and to make one galactic orbit takes the Sun in the region of 250 million years.

galaxy (*plural* **galaxies**) the name given to the band of stars, numbering 10^{11} bodies, which includes the SUN, and is alternatively called the Milky Way. The galaxy has a spiral structure and is approximately 10^5 LIGHT YEARS across.

gale a wind of force 8 on the BEAUFORT SCALE (more than 30 knots).

galena the commonest lead ore, occurring as lead sulphide (PbS), often in the form of grey cubic crystals and frequently twinned (*see* TWINNING). It is often found in association with blende (zinc sulphide), and silver sulphide may be present in quantities sufficient to warrant processing for silver. It is commonly formed by METASOMATISM or hydrothermal activity.

gall bladder *see* **bile**.

galvanic cell an energy-producing electrochemical cell.

galvanizing the process by which one type of metal is coated with a thin layer of another, more reactive metal. Galvanizing is performed for the purpose of protection, as the more reactive metal coating will corrode before the underlying metal. For instance, sheets of iron and steel are often coated with the more reactive metal, zinc. Even when the zinc coating becomes damaged, the underlying iron or steel will be protected.

gamete the reproductive cell of an organism. Gametes can be either male or female, and these specialized cells are HAPLOID in number but unite during fertilization, producing a DIPLOID ZYGOTE that later develops into a new organism. In higher animals, the male and female cells are called sperm and ova respectively, whereas in higher plants they are known as pollen grains and egg cells respectively. In some organisms there is essentially one type of gamete that is capable of developing into a new individual without fertilization. These gametes are usually diploid, as in the case of certain lower plant groups, e.g. many forms of algae.

gamma ray a type of ELECTROMAGNETIC radiation

released during the radioactive decay of certain nuclei. The rays released are the most penetrative of all radiations, requiring about twenty millimetres of lead to stop them. The gamma rays are useful for sterilizing substances and in the treatment of cancer. They have the shortest wavelength of any wave in the electromagnetic spectrum, i.e. 10^{-10} to 10^{-12} metres.

ganglion (plural **ganglia**) a mass of nervous tissue that contains cell bodies (the part of the nerve cell with the nucleus) and SYNAPSES. Ganglia comprise part of the central nervous system (CNS) in invertebrates, occurring along the nerve cords, but in vertebrates they occur outside the CNS in the main.

garnet (the) garnet (family) is a widely occurring group of minerals found particularly in METAMORPHIC rocks (SCHISTS and GNEISSES) but also in igneous rocks and reworked deposits such as beach sands and PLACERS. They are silicates of various di- and trivalent metals, with the general formula $R^{2+}{}_3 R^{3+}{}_2 (SiO_4)_3$, where R^{2+} can be calcium, magnesium, iron or manganese, and R^{3+} can be iron, aluminium, chromium or titanium. The metal ions lie between, and bond together, the SiO_4 groups. Garnets vary enormously in colour, depending on composition, and include a number of semi-precious varieties: pyrope (red, the MgAl form); grossular or cinnamon stone (light cinnamon, the CaAl form); and almandine (brownish-red, the FeAl form). Non-gem quality garnets are used for abrasives.

gas the fluid state of matter capable of indefinite expansion in every direction, due to the relatively few bonds between the atoms or molecules present in the gas. If heat ENERGY is supplied to a gas, it expands to the limits of its containing vessel, exerting a pressure on this vessel that in turn exerts a force back onto the gas.

gas analysis the analysis of a mixture of gases by various means. By putting a measured quantity of gas in contact with various reagents, certain components can be measured after each phase of absorption. Carbon dioxide is absorbed by potassium hydroxide, carbon monoxide in acid or alkaline CuCl, oxygen in alkaline pyrogallol (trihydroxybenzene— $C_6H_3(OH)_3$), and so on. Additional procedures can also be adopted for gas analysis, including TITRATION, measurement of spectra (infrared or ultraviolet), or CHROMATOGRAPHY.

gas laws the rules that relate to the pressure, temperature and volume of an ideal gas, allowing useful information about a gas to be gained by calculation instead of by experimentation. The laws are termed BOYLE'S LAW, CHARLES' LAW, and the pressure law. The pressure law states that, when a gas is kept in a constant volume, the pressure of that gas will be directly proportional to the temperature. All three laws can be combined in an equation known as the universal gas equation, which allows gases to be compared under different temperatures and pressures, i.e.

$$pV = nRT,$$

where p, V and T relate to pressure, volume and temperature respectively, n is the quantity of gas under investigation, and R is the universal molar gas constant, which has the value of 8.314 $JK^{-1}mol^{-1}$.

gas show a surface expression of natural gas escaping from reservoirs underground.

gastrin a hormone, secreted in the stomach, that

stimulates the gastric glands of the stomach to produce gastric juice. The presence of food in the stomach is the initial stimulus, and the gastrin controls the digestive process. Gastric juice is a mixture of hydrochloric acid, certain salts and some ENZYMES, e.g. PEPSIN, which catalyse the breakdown of protein.

Geiger tube *or* **Geiger-Müller tube** an instrument, named after the German physicist Hans Geiger (1882-1945), that can detect and measure radiation. The tube contains an inner electrode and a cylindrical outer electrode filled with a gas at low pressure. Any radiation enters the tube through a mica window, causing an electrical pulse to travel between the electrodes. These pulses are detectable when the Geiger tube is connected to an electronic circuit, called a scaler, which records the total radiation in the area in a given time.

gel a jelly-like material resulting from the setting of a COLLOIDal solution. The VISCOSITY is often such that the solution may have properties more like solids than liquids.

gem gravels sediments containing significant amounts of gem minerals. The gravels are formed by the breakdown and transport of the rocks in which the gems originated.and are really a type of PLACER, without gold but containing GARNET and rubies, etc. The heavy and more stable gem minerals resist weathering and washing away, unlike the lighter components of the rocks, and a concentration is thus produced.

gene the chemically complex unit of heredity, found at a specific location on a CHROMOSOME, that is responsible for the transmission of information from one generation to the next. Each gene contributes to a particular characteristic of the organism, and gene size varies according to the characteristic that it codes for. For example, the gene that codes for the HORMONE called insulin, consists of 1700 BASE PAIRS on a DNA molecule.

gene cloning a method of GENETIC ENGINEERING whereby specific genes are extracted from host DNA and introduced into the cell of another host by means of a plasmid VECTOR. All the descendants of the genetically transformed host cell will produce a copy of the gene. The transformed gene is thus said to have been cloned (*see also* CLONE).

gene flow the transfer of genes between populations via the GAMETES. Gene flow enhances variation in a population as it can lead to a change in the frequency of ALLELES present within that population. This in turn is a factor that contributes towards EVOLUTION, as the alleles affect the characteristics of an organism. Therefore, gene flow can be advantageous, as it can help an organism inherit new characteristics that may be beneficial to its survival.

generator *see* **dynamo**.

genetic adaptation *see* **adaptation**.

genetic engineering the branch of biology that involves the artificial modification of an organism's genetic make-up. The term covers a wide range of techniques, including selective plant and animal breeding, but it is especially associated with two particular techniques:

(1) The transfer of DNA from one organism to a different organism in which it would not normally occur. For example, the gene that codes for the

human hormone, insulin, has been successfully incorporated into the GENOME of bacterial cells, and the bacteria produce insulin.

(2) Recombination of DNA between different species in the hope of producing an entirely new species. For instance, cells of the potato and tomato plants, which have had their cell walls removed, have been successfully cultured and made to fuse together using a variety of experimental procedures. Such cells can grow successfully and develop into a new species of plant that has been called the pomato. Although crossing the species barrier is an important breakthrough in the field of genetic engineering, there are strict governmental regulations regarding the release of such species into the environment since the consequences cannot be predicted.

genetic recombination the exchange of genetic material during meiosis, with the effect that the resultant GAMETEs have gene combinations that are not present in either parent. This rearrangement of genes allows for variability in a species, and in each generation an almost infinite variety of new combinations of ALLELES of different genes are created. Such novel combinations of genes can confer enormous benefits to an organism when conditions change. For example, only a tiny number of a population of locusts have specific combinations of genes that enable them to survive potent pesticides. When such insects reproduce, they produce resistant populations—a major problem in the world of agriculture.

genome the total genetic information stored in the CHROMOSOMES of an organism. The number of chromosomes is characteristic of that particular species of organism. For instance, a man has 23 pairs of HOMOLOGOUS CHROMOSOMES (containing approximately 50,000 genes), domestic dogs have 39 pairs, and domestic cats have 19 pairs. In each case, one pair of chromosomes constitute the SEX CHROMOSOMES, and the remaining pairs are the AUTOSOMES.

genotype the specific versions of the GENES in an individual's genetic make-up. For instance, there are three possible genotypes for the human albino gene, and it has two allelic forms, dominant A and recessive a. Thus, the three possible genotypes are:

(1) AA (homozygous dominant)

(2) aa (homozygous recessive)

(3) Aa (heterozygous).

geocentric a descriptive term meaning any system that has the centre of the Earth as its reference point.

geocentric parallax in astronomy, the movement of an observer due to the Earth's rotation, causing an apparent positional change of a heavenly body. This is the geocentric parallax.

geochemistry the study of the chemical constitution of the Earth (or solid bodies within the SOLAR SYSTEM), including the abundance and distribution of chemical elements and their isotopes.

geochronology (*see also* ISOTOPE GEOCHEMISTRY) the study of time on a geologic scale through the use of *absolute* and *relative age-dating* methods. Relative age-dating deals with the study and use of FOSSILS and SEDIMENTS to put rock successions and sequence in order. Absolute methods provide an actual age for a rock, using radioactive elements whose decay rates (*see* HALF-LIFE) are known, enabling the necessary calculations to be made.

geode a rounded "rock" (similar to a large potato in size and appearance), which on examination proves to be hollow and contains mineral crystals growing from the wall to the centre. Their unimpeded growth into the cavity often produces crystals with perfect HABIT, which are valued by collectors.

geodesic in geometry, the shortest path between two points.

geodesy essentially a combination of mathematics and surveying, involving the measurement of the shape of the Earth (or large areas of it).

geology the scientific study of planet earth. This includes geochemistry, petrology, mineralogy, geophysics, palaeontology, stratigraphy, physical and economic geology.

geometry a major branch of the mathematical sciences, which involves the study of the relative properties of various shapes. For example, the calculations used to determine the size of the angles and the area of a triangle.

geophysics the study of processes within, and the properties of, the Earth. Included within the discipline are SEISMOLOGY, magnetism, gravity, HYDROLOGY, GEOCHRONOLOGY and studies of heat flow.

geostationary a term referring to an orbit around the Earth, in which a satellite stays in the same position relative to the Earth because it rotates at the same speed as the planet.

geostrophic force (*see* CORIOLIS FORCE) a theoretical force used to accommodate for the EARTH's rotation, which produces a change in wind direction relative to the surface of the Earth.

geosyncline a largely defunct term for a very large downward warping in the Earth's crust that formed a sedimentary basin. The idea was that the geosyncline subsided and was gradually filled with sediment from neighbouring land masses, and with the products of volcanic eruptions and INTRUSIONS. These rocks were subsequently deformed, folded and metamorphosed, although the mechanisms were never truly understood. The whole theory has now been overtaken by the concept of PLATE TECTONICS, although some of the terminology remains.

geothermal gradient the increase of temperature with depth, into the EARTH. The gradient varies with location, from continents to oceans, and depends greatly upon the tectonic and volcanic activity of a region. In general the gradient is within the range 15° to 40°Ckm^{-1} .

geotropism a growth movement, exhibited by plants in response to the force exerted by GRAVITY. Plant roots are termed positively geotropic since they grow downwards, whereas plant shoots generally grow upwards (towards sunlight) thus displaying negative geotropism.

germination the start of growth in a dormant structure, e.g. a seed or spore. Various factors can break seed dormancy, such as specific temperatures, exposure to light, or rupture of the seed coat, all of which depend on the species from which the seed is derived.

gestation period the period from conception to birth in mammals, which is characteristic of the species concerned. For instance, dogs have a gestation period that on average is 63 days, whereas that of the blue whale is 11 months.

geyser a small fissure or opening in the Earth's surface, connected to a hot spring at depth, from which a column of boiling water and steam is ejected periodically. The mechanism of eruption is created by hot rocks heating water to boiling point at the base of the column before the top. Vapour bubbles rise through the column with considerable force, pushing out water and steam at the top. This reduces pressure at the base of the column, and boiling continues. Minerals dissolved in the water (calcium carbonate, or silica) are deposited around the mouth of the geyser.

giant star a star in a spectral class (*see* SPECTRAL TYPES) which is brighter than the main stars in the class.

gibberellin *see* **hormone**.

Gibbs free energy *see* **free energy**.

Gibbs phase rule *see* **phase rule**.

glacial action all processes related to the action of a glacier, including accumulation of crushed rock fragments and the physical actions, e.g. grinding, scouring and polishing, which are all due to the incorporation of rocks into the ice.

glacial deposits all deposits formed by some action of glaciers, e.g. BOULDER CLAY, sands and gravels occurring as OUTWASH FANS, and deposits in the form of DRUMLINS and ESKERS.

glacial erosion the removal and wearing down of rock by glaciers and associated streams (of meltwater).

glaciation a term meaning ice-age, with all its effects, processes and products. The most recent is associated with the Pleistocene (*see* APPENDIX 5), but the rock record indicates older glaciations from the

Precambrian and Permo-Carboniferous, and other periods in geological history.

glacier an ice mass of enormous size, usually moving. Three kinds are specified on occurrence and morphology, namely PIEDMONT GLACIERS, valley glaciers, and ice-sheets (e.g. Greenland). An alternative classification is based on temperature and covers polar glaciers (e.g. in the Antarctic), which have very low temperatures and which move slowly, primarily by deformation within the body of ice; temperate glaciers (e.g. in the Alps), where movement occurs largely by slip at the base; and subpolar glaciers (e.g. Spitzbergen), which are a mix of the two types.

globular cluster a spherical arrangement of stars, closely packed together, and containing up to many millions of stars. About one hundred clusters occur in the Milky Way, throughout the GALACTIC HALO.

globule a nebula, made up of opaque dust and gas, that denotes an early stage of star formation.

gluconeogenesis a major metabolic pathway occuring predominantly in the liver, which synthesizes glucose from non-carbohydrate precursors in conditions of starvation. Glucose is required by red blood cells and is the primary energy source of the brain. However, the glucose reserves present in body fluids are sufficient to meet the body's needs for only about one day. Therefore, gluconeogenesis is very important during longer periods of starvation or during periods of intense muscle exercise. There are three major non-carbohydrate classes that serve as raw materials for gluconeogenesis:

(1) GLYCEROL—derived from fat hydrolysis.

(2) AMINO ACIDS—derived from protein degradation

during starvation and from proteins in the diet.
(3) Lactate (lactic acid)—formed by actively contracting muscle when there is an insufficient supply of oxygen. (It is also produced by red blood cells).

glucose the most abundant naturally occurring sugar, which has the general formula $C_6H_{12}O_6$. Glucose is distributed widely in plants and animals and is an important primary energy source, although it is usually converted into polysaccharide carbohydrates, which serve as long-term energy sources. The storage polymers of plants and animals are starch and GLYCOGEN respectively. Other polysaccharides of glucose include chitin and cellulose, which have a structural role and also provide strength.

glycerol a viscous, sweet-smelling alcohol, which has the chemical formula $HOCH_2CH(OH)CH_2OH$. Glucose is widely distributed in plants and animals as it is a component of stored fats. During metabolism, stored fats break down to form the original reactants, glycerol and FATTY ACIDS, while a large amount of energy is released. Glycerol is used commercially to manufacture a wide range of products, including explosives, resins, toilet preparations and foodstuffs.

glycogen, *often called* **animal starch**, a polySACCHARIDE of GLUCOSE units that occurs in animal cells (especially the muscle and liver) and acts as a store of energy released upon hydrolysis. Glycogen is also found in some fungi.

glycolysis a major metabolic process, occurring in the CYTOPLASM of virtually all living cells, where the breakdown of glucose into simple molecules generates energy in the form of ATP. Each 6-carbon glucose molecule is converted into two 3-carbon pyruvate molecules ($CH_2COCOOH$) in a sequence of ten reactions, giving a net gain of two ATP molecules. The glycolytic pathway is regulated by several ENZYMES. Although the reactions converting glucose to pyruvate are very similar in all living organisms, the fate of pyruvate is variable. In AEROBIC organisms, pyruvate enters the MITOCHONDRIA, where it is completely oxidized to CO_2 and H_2O in a process known as Kreb's cycle (or CITRIC ACID CYCLE). This cycle, together with glycolysis, liberates 38 molecules of ATP per glucose molecule. However, if there is an insufficient supply of oxygen, e.g. in an actively contracting muscle, FERMENTATION occurs and pyruvate is converted into lactic acid, liberating only 2 ATP molecules per glucose molecule. In some ANAEROBIC organisms, such as YEAST, pyruvate is converted into the alcohol ETHANOL during fermentation, again yielding only 2 ATP molecules. If a cell requires energy, or certain intermediates of the pathway are required for the synthesis of new cellular components, glycolysis proceeds, provided that glucose levels in the blood are abundant. However, when blood-glucose levels are low, e.g. during starvation, glycolysis is inhibited and instead GLUCONEOGENESIS occurs. Glycolysis and gluconoegenesis are reciprocally regulated so that when one process is relatively inactive, the other is highly active.

glycoprotein a protein that has a sugar residue linked to it by covalent bonding (*see* GLYCOSYLATION).

glycosylation the process by which a carbohydrate is added to an organic compound, for example a protein. Glycosylation may occur in the ENDOPLASMIC

RETICULUM or the GOLGI APPARATUS of cells and plays an important role in regulating protein activity.

Gmelin test a test used to determine the presence of BILE pigments, which results in the formation of coloured oxidation products.

gneiss a coarse-grained METAMORPHIC rock formed during high-grade REGIONAL METAMORPHISM. Gneisses typically contain QUARTZ and FELDSPAR and show banding due to segregation of light and dark minerals, the dark minerals being BIOTITE, hornblende (*see* AMPHIBOLES) or PYROXENES. Gneisses derived from SEDIMENTARY rocks are termed paragneisses, and those of IGNEOUS origin, orthogneisses.

Golgi apparatus a system of ORGANELLES within the cells of organisms, comprising stacks of flattened sacs that act as the assembly point for the modification, sorting and packaging of large molecules; proteins, for example, undergo GLYCOSYLATION here. Numerous small membrane-bound vesicles surround the Golgi apparatus, and these are thought to transport the modified macromolecules from the Golgi apparatus to the different compartments of the cell. It is named after the Italian physician Camillo Golgi (1844–1926) who discovered its existence.

gonads the reproductive organs of animals, which produce the GAMETES and certain HORMONES. The male and female organs are known as the testes and the ovaries respectively.

Gondwanaland the massive, hypothetical, continent in the southern hemisphere that gave rise to parts of the present Africa, South America, India, Australia, New Zealand and Antarctica. Their connection at one time is postulated as a reason for the occurrence of widely separated but similar groups of plants and animals.

graben an area of land that is bounded by two or more NORMAL FAULTS with opposite senses of movement, resulting in a central block that falls relative to either side. Such features that extend for long distances are called rifts.

graded bedding a primary feature in sedimentary rocks, which shows a gradation in grain size from the base to the top of the bed. At the base is coarse-grained sand or pebbles, and rising through the bed the size diminishes through fine sand, to silt and clays on top (*see also* FLYSCH).

gradient a measure of the steepness of a sloping line. A straight line has the equation $y = mx + c$, where x and y are the co-ordinates, c is a constant, and m is the gradient. The steepness of a point on a curve is the gradient of its tangent, which is a straight line drawn to the curve at this point.

gradient wind a wind that is the resultant of all the forces that act upon moving air—the pressure gradient, CORIOLIS FORCE, and centrifugal force.

grain shape, grain size *see* **particle shape, particle size**.

gram (g) the basic unit of mass. There are approximately 28g in an ounce, and precisely 1000g in a kilogram.

granite a coarse-grained and commonly occurring IGNEOUS rock containing QUARTZ, alkali FELDSPAR and MICA (usually BIOTITE). Other minerals may also be present, including AMPHIBOLES and oxide minerals. Granites can be formed by several processes, including melting of old continental crust and FRACTIONAL CRYSTALLIZATION of BASALT MAGMA.

graph a diagram that represents the relationship between two or more quantities, using dots, lines, bars or curves.

graphite a soft, black, hexagonal variety of CARBON with a very high melting point that makes it chemically inert. It is a giant structure comprising a series of planes. In any one plane the bonds between the atoms are strong, but bonds between the atoms of different planes are weak. These properties account for the fact that graphite is a good lubricant and conductor of electricity. It is used in the making of pencils and electrodes.

graptolite an extinct marine organism that secreted CHITINOUS tubes attached to branches, forming small (several cm) colonies. They first appeared in the Middle Cambrian and form useful INDEX FOSSILS for the Ordovician and Silurian.

gravimetric analysis a method of chemical analysis in which elements are precipitated from a solution in the form of a known compound to be weighed and from which the quantity of the required element can be calculated.

gravity the attractive force that the earth exerts on any body that has mass, tending to cause the body to accelerate towards it. Other planets also exert a force of gravity, but the force is different from that exerted by the earth since it depends on the planet's mass and diameter. The true WEIGHT of any object on earth is really equal to the object's MASS (m) multiplied by the ACCELERATION due to gravity (g), which is 9.8 ms^{-2}. Therefore, although weight and mass are often used synonymously, they are different for scientific purposes. For example, a man with a mass of 80kg will weigh 784 newtons (N) on earth, but on the moon he would weigh only 130N since the force of gravity on the moon is only 1/6th of that on earth. However, his mass is still 80kg and remains constant throughout the universe.

green flash an astronomical occurrence in clear atmospheres, when a bright sun disappears below the horizon as a circle of light and green is the last refracted colour to be seen. Blue, although it is refracted more, is dispersed.

greenhouse effect the phenomenon whereby the earth's surface is warmed by solar radiation. Most of the solar radiation from the sun is absorbed by the earth's surface, which in turn re-emits it as INFRARED RADIATION. However, this radiation becomes trapped in the earth's atmosphere by carbon dioxide (CO_2), water vapour and OZONE, as well as by clouds, and is re-radiated back to earth, causing a rise in global temperature. The concentration of CO_2 in the atmosphere is rising steadily because of mankind's activities (e.g. deforestation and the burning of FOSSIL FUELS), and it is estimated that it will cause the global temperature to rise 1.5-4.5°C in the next fifty years. Such a rise in temperature would be enough to melt a significant amount of polar and other ice, causing the sea level to rise by perhaps as much as a few metres. This could have disastrous consequences for coastal areas, in particular, major port cities like New York.

greywacke a type of sandstone that is poorly sorted, with a large range in grain size and a high percentage (~15 per cent) of clay minerals that form the matrix. The grains present include QUARTZ, MICA, FELDSPAR, rock fragments and possibly minerals rich in iron and magnesium. These features indicate that greywackes are formed in regions of intense erosion with rapid transport and deposition.

groundmass the term applied to the fine-grained matrix of IGNEOUS ROCKS in which there may be larger crystals. The groundmass represents a second phase of crystallization characterized by rapid cooling with the generation of minute crystals and/ or glass. In lavas, the fine groundmass is formed upon eruption onto the surface.

groundwater water contained in the voids within rocks. It usually excludes water moving between the surface and the WATER TABLE (*vadose* water) but may be METEORIC or JUVENILE water.

group the vertical columns of elements in the PERIODIC TABLE. Each group contains elements that have similar properties, including the same number of electrons in the outer energy level (shell). This number is represented by the group number. For example, the alkali metals in group 1 all have one electron in the outer shell, whereas the HALOGENS in group 7 all have 7 electrons in the outer shell.

guanine ($C_5H_5N_5O$) a nitrogenous base component of the nucleic acids, DNA and RNA. It has a PURINE structure with a pair of fused HETEROCYCLIC rings, which contain nitrogen in addition to carbon. In both DNA and RNA, guanine always BASE PAIRS with CYTOSINE, which has a pyrimidine structure. Guanine is also present in many other biologically important molecules.

gum arabic *or* **acacia gum** a white, water-soluble powder, which in natural form is obtained from some varieties of acacia trees. It is a complex polySACCHARIDE and is used widely in pharmacy as an emulsifier (*see* EMULSION), and as an adhesive. It is also used in the food industry as an emulsifier and inhibits sugar crystallization.

gypsum an EVAPORITE mineral, $CaSO_4.2H_2O$, that is highly insoluble and thus the first to precipitate out of sea water, to be followed by ANHYDRITE and HALITE (*see also* ALABASTER). It commonly forms crystals, often twinned (*see* TWINNING) and is very soft (2 on MOHS' SCALE). It is important industrially, being used in cements, fertilizer, fillers, and plaster products.

H

Haber process the industrial process for the production of ammonia (NH_3) by the direct combination of nitrogen and hydrogen in the presence of an iron CATALYST. The process gives a maximum yield (40 per cent) using relatively low temperatures and high pressures. The Haber process is important in industrial chemistry since it is the most economic way to produce ammonia, from which fertilizers are made.

habit a term denoting the characteristic external shapes of crystals due to the number, shape, size, and orientation of the crystal faces. The extent to which perfect habit may be approached depends upon conditions during formation, speed of growth, impurities, etc. Individual crystals of particular minerals may show typical growths. Habits include tabular, fibrous, prismatic (elongated) and ACICULAR (needle-shaped).

habitat the place where a plant or animal normally lives, specified by particular features, e.g. rivers, ponds, sea shore.

hade when dealing with a non-vertical FAULT plane, the hade is the angle between that plane and the vertical.

hadron *see* **elementary particle**.

haem *see* **myoglobin**

haemoglobin an iron-containing red pigment that is found within the red blood cells (or ERYTHROCYTES) of vertebrates and that is responsible for the transport of oxygen around the body. In actively metabolizing tissue, e.g. the muscles, haemoglobin exchanges oxygen for carbon dioxide (CO_2), which is then carried in the blood back to the heart and pumped to the lungs, where the haemoglobin loses the CO_2 and regains oxygen.

haemophilia a genetic disorder affecting the blood, in which the lack of a vital BLOOD CLOTTING factor causes abnormally delayed clotting. Haemophilia is exhibited almost exclusively by males, who receive the defective gene from their mothers. A haemophilic female can only arise if a haemophilic male marries a female carrying the gene (extremely rare). There is no known cure for haemophilia, and, when injured, haemophiliacs must rely on blood transfusion to replace the blood loss, which is considerably greater than that lost by a normal individual.

haemostasis *see* **blood clotting**.

hail hard balls or pellets of ice usually associated with CUMULONIMBUS cloud. The hail(stones) form by rain drops being taken into higher, colder regions. Upon falling, the hailstone grows by accumulations of layers due to condensation of moist air on the cold hail.

half-cell as implied, half of an electrolytic cell (*see* ELECTROLYTE). The cell comprises an ELECTRODE dipped into an electrolyte, and the potential of this system is measured against a hydrogen electrode that is given a potential of zero. A hydrogen electrode consists of hydrogen bubbling over a platinum electrode that is covered by dilute acid to produce a standard concentration of hydrogen IONS.

half-life (t) the time taken for a radioactive ISOTOPE to lose exactly half of its RADIOACTIVITY. The half-life is constant for a particular isotope, varying from a fraction of a second to millions of years, and is best determined by using a Geiger-counter (GEIGER TUBE). For instance, if an isotope has a half-life of one minute, then the radioactive count will fall by one half in one minute, by one quarter in two minutes, by one eighth in three minutes, and so on.

halide a compound consisting of a HALOGEN and another element. Halides are ionically bonded when formed by electropositive metals, e.g. sodium bromide (NaBr). Halides formed by less electropositive metals and non-metals have COVALENT BONDS.

halite *or* **rock salt** (NaCl) a common mineral in EVAPORITE deposits and often associated with GYPSUM and ANHYDRITE. Considerable thicknesses may occur and can be mined or extracted as brine. Large masses can form "plugs," which arch into overlying sedimentary rocks (*see* DIAPIR) creating oil traps.

halo a bright ring or rings that may be seen around the Sun or Moon. It is due to refraction of light by the crystals in high CIRRUS cloud.

halogen any of five elements, found in group 7 of the PERIODIC TABLE, that are the extreme form of the non-metals. They exhibit typical non-metal characteristics, existing as COVALENTly BONDed diatomic molecules, e.g. F-F (a fluorine molecule). At room temperature, fluorine and chlorine are gases, bromine is a volatile liquid, and iodine is a volatile solid. The halogens are found in nature as negative IONS in sea water and as salt deposits from dried-up seas.

halophilic bacteria bacteria that can tolerate salt and live in the surface layers of the sea. They are instrumental in various biochemical cycles, including the nitrogen and carbon cycles.

halophyte a plant that can tolerate a high level of salt in the soil. Such conditions occur in salt marshes, tidal river estuaries (and on motorway verges and central reservations!). A typical species is rice grass (*Spartina*).

hanging valley a tributary valley situated above a major valley, possibly with a waterfall. It is formed by the greater erosion of the main (trunk) valley by its GLACIER. The main glacier also cuts off the ends of the land between adjacent hanging valleys, creating "truncated spurs."

haploid this term describes a cell nucleus or an organism that possesses only half the normal number of CHROMOSOMES, i.e. a single set of unpaired chromosomes. This is characteristic of the GAMETES and is important at fertilization as it ensures the DIPLOID chromosome number is restored. For example, in a man there are 23 pairs of chromosomes per somatic cell, which is the diploid number, but the gametes possess 23 single chromosomes, which is the haploid number.

hardness of minerals (*see* **mineralogy**)

hard water water that does not readily form a lather with SOAP. This is due to dissolved compounds of calcium, magnesium and iron. Use of soap produces a scum which is the result of a reaction between the FATTY ACIDS of the soap and the metal ions. The scum is made up of SALTS, which when removed render the water soft. There are two types of hardness. Temporary hardness is created by water passing over carbonate rocks (e.g. limestone or chalk), producing hydrogen carbonates of the metals, which dissolve before the water reaches the mains supply. Boiling the water decomposes the hydrogen carbonates into carbonates (producing kettle fur), and the water becomes soft. Permanent hardness in water is due to metal sulphates, which can be removed by the addition of sodium carbonate. ZEOLITES will remove both types of hardness.

Hardy-Weinberg ratio a law that states that in a large, randomly breeding population, the genetic and allelic frequencies will remain constant from generation to generation. For example, a particular GENE in a population may have a number of ALLELES, one of which is dominant, but this does not necessarily mean that it occurs at a higher frequency than the recessive alleles. If the gene has two alleles, B (dominant) and b (recessive), present at frequencies x and y respectively, then the proportion of the genotypic frequencies would be:

$$BB \quad Bb \quad bb$$
$$x^2 \quad 2xy \quad y^2 = 1.0$$

There are certain conditions of stability that must be met for such a genetic equilibrium to occur:

(1) The population must be large, so that allelic frequencies could not be altered by chance alone.

(2) There must be no mutation, or it must occur in equilibrium.

(3) There must be no immigration or emigration, which would alter the genetic frequencies in question.

(4) Mating and reproductive success must be completely random with respect to GENOTYPE. If all conditions of the Hardy-Weinberg law are met, EVOLUTION could not occur as allelic frequencies would not change. However, evolution does occur

171

because the conditions are never entirely met. For example, the condition of random mating is probably never met in any real population since an organism's genotype almost always influences its choice of a mate and the physical efficiency and frequency of its mating. The condition regarding mutation is probably never met either, since mutations are always occurring, resulting in a slow shift in the allelic frequencies in the population, with the more mutable alleles tending to become less frequent.

Harvard classification a method for classifying the spectra of stars.

haze the reduction of visibility in atmospheric conditions due to the dispersion of light by very small dust or smoke particles.

head a superficial deposit comprising poorly sorted, angular fragments of rock that form as a result of freezing and thawing and are then deposited by gelifluction (waterlogged debris sliding downhill). Head is found in valleys that were near the edge of an ice sheet.

heart a hollow, muscular organ that acts as a pump to circulate blood throughout the body. The heart lies in the middle of the chest cavity between the two lungs. It is divided into four chambers, known as the right and left ATRIA, and the right and the left VENTRICLES. In normal persons there is no communication between the right side and the left side of the heart, thus the two sides act as independent pumps, which are connected in series. Starting from the left ventricle, the flow of blood is as follows:
(1) Left ventricle contracts and oxygenated blood is pushed into the AORTA under pressure.

(2) Aorta divides into numerous ARTERIES to supply blood to all parts of the body.

(3) Deoxygenated blood returning from the body is carried by small VEINS, which eventually join up to form two large veins, called the superior VENA CAVA and the inferior vena cava.

(4) These two large veins empty into the right atrium.

(5) The blood passes from the right atrium to the right ventricle via a VALVE.

(6) The right ventricle contracts, pushing blood under pressure into the PULMONARY ARTERY.

(7) The pulmonary artery branches into two, carrying blood to both the right and left lungs.

(8) Within the lungs, gas exchange occurs—carbon dioxide is expelled, and the blood is oxygenated (*see* HAEMOGLOBIN).

(9) The blood flows from the left atrium via a valve into the left ventricle.

heat ENERGY produced by molecular agitation.

heat capacity (C) the quantity of heat required by a substance or material that will raise its temperature by one degree KELVIN (or one Celsius). Thus, heat capacity is measured in joules per Kelvin (JK^{-1}) or joules per Celsius (JC^{-1}). The molar heat capacity of a substance is the heat required that will raise the temperature of one MOLE of the substance by one degree, and the specific heat capacity is the heat capacity per kilogram ($JK^{-1}kg^{-1}$) or gram ($JK^{-1}g^{-1}$).

heat exhaustion a physical state experienced by warm-blooded animals whereby the body's normal cooling processes fail to operate as a result of increasing environmental temperature. Instead, the body's metabolic rate increases, raising the body temperature higher, which in turn raises the metabolic rate even higher, and so on. The symptoms of heat exhaustion are cramp and dizziness, and death ensues when the body temperature reaches about 42°C, which is the upper lethal temperature for the average human being.

heat of combustion the amount of heat generated when one MOLE of a substance is burned in oxygen.

heat of dissociation the amount of heat required to dissociate (*see* DISSOCIATION) one MOLE of a compound.

heat of formation the heat required or given out when one MOLE of a substance is formed from its elements (at one atmosphere and usually 298K).

heat of reaction the amount of heat absorbed or given out for each MOLE of the reactants. If the reaction is EXOTHERMIC, the convention is to specify the quantity as a negative figure (in kilojoules).

heat of solution the amount of heat absorbed or given out when one MOLE of a substance is dissolved in water.

heave the horizontal movement component in a dip slip FAULT (i.e. where the movement is parallel to the dip of the fault plane).

hela cell a particular cell variety, discovered in a woman with cervical carcinoma (a form of CANCER) in 1951. These transformed cells are immortal and are used in laboratories worldwide for research purposes.

helical rising *or* **setting** when a star or planet rises (or sets) at the same time as the Sun.

heliometer an apparatus for measuring the diameter of the SUN or the angular distance between two objects that are very close together.

heliostat an astronomical instrument, similar to the

COELOSTAT, that enables study (both by photography and spectroscopy) of the Sun to be made.

helium an INERT GAS with a stable NUCLEUS identical to an α particle (*see* ALPHA DECAY). It occurs naturally in small quantities, and, being nonflammable and light, is used for airships and balloons. Helium liquefies below 4K and is the commonest coolant used in CRYOGENICS.

helium stars SPECTRAL TYPE B stars that show dark lines due mainly to helium.

helix a curve in the form of a spiral, which encircles the surface of a cone or cylinder at a constant angle.

hemisphere a half sphere, formed when a plane is passed through the middle of a sphere, cutting it in two.

Henry's law this states that the amount of gas dissolved in a given quantity of liquid is directly proportional to the gas pressure above the solution. Henry's law only applies to oxygen and inert gases, i.e. GROUP 8 of the periodic table. It does not apply to carbon dioxide or ammonia, for example, as both will react with water. Henry's law is represented by the following equation: $c = kP$, where c is MOLARITY, k is the absorption coefficient, and P is the pressure. The temperature must be constant as heat decreases the solubilities of gases in liquids but increases the solubilities of liquids in liquids.

hepatitis a serious disease of the liver, of which there are two types:

(1) Hepatitis A (infectious hepatitis) caused by the hepatitis A virus

(2) Hepatitis B (serum hepatitis) caused by the hepatitis B virus

Both diseases share the same symptoms of fever,

nausea and JAUNDICE, but they are transmitted by different routes. Hepatitis A is spread by the oral-faecal route and occurs in people who have poor sanitation and personal hygiene. The virus can be transmitted from person to person in contaminated food or drinking water. Most people exposed to the disease can be protected by PASSIVE immunization, which involves the administration of purified ANTI-BODIES from a previously infected individual who has recovered. Hepatitis B is spread through blood products, contaminated syringes and instruments. Susceptible groups include those who require blood or blood products, e.g. haemophiliacs (although any donated blood is normally screened for hepatitis). A significant percentage of hepatitis B sufferers develop cancer, and the virus is thought to be a contributory factor.

heptagon a polygon with seven sides, the interior angles of which add together to give 900°. If the heptagon is regular, then all the sides are equal.

hertz (Hz) the unit of FREQUENCY, equivalent to one cycle per second. For example, if an OSCILLATION has a frequency of 6Hz, this means that six complete cycles occur in one second.

Hertzsprung-Russell diagram a graphical means of correlating star data, in which LUMINOSITY is plotted against SPECTRAL TYPE, first used by the American astronomer Henry Norris Russell (1877–1957) in 1913, based on studies previously carried out by the Danish astronomer Ejnar Hertzsprung (1873–1967). These features relate to, and the graph is basically a plot of, total energy against surface temperature. The outcome is that most stars form a band across the graph from hot, bright, blue-white stars to cool, dim, red stars at the two extremes. This band is the MAIN SEQUENCE containing normal stars burning hydrogen. As stars evolve, they move away from the main sequence, firstly as RED GIANTS. The diagram is fundamental to theories of STELLAR EVOLUTION.

Hess's Law states that irrespective of the intermediate stages, the overall heat change in a chemical reaction depends only on the beginning and final states.

heterocyclic compounds organic compounds forming a ring structure with the additional elements, e.g. oxygen, hydrogen, nitrogen and sulphur.

heterozygote an organism having two different ALLELES of the GENE in question in all somatic cells. For instance, if gene B has two allelic forms, B and b, then the heterozygote will contain both alleles, i.e. Bb, at the appropriate location on a pair of HOMOLOGOUS CHROMOSOMES. Heterozygotes can thus produce two kinds of GAMETES, B and b. One allele of a heterozygote is usually dominant, and the other is usually recessive. The dominant allele is the one that is expressed phenotypically, because it masks the expression of the recessive allele. Dominant alleles are usually denoted by capital letters, while recessive alleles are denoted by lower case letters.

hexagon a polygon with six sides. If the hexagon is regular, then each of the interior angles is 120°.

hiatus a break in a succession of sedimentary rocks due to erosion or non-deposition.

Hill reaction the light-dependent stage of PHOTOSYNTHESIS, in which illuminated CHLOROPLASTS initiate the photochemical splitting of water. This produces hydrogen atoms (two per water molecule), which are used to reduce carbon dioxide with the formation of carbohydrate in the dark stage of photosynthesis. The light stage also produces ATP, which provides the energy required for carbohydrate synthesis.

histogram a graph that represents the relationship between two variables using parallel bars, but it differs from a bar chart in that the frequency is not represented by the bar height, but by the bar area.

histamine an AMINE released in the body during allergic reactions, and in injured tissues. Release causes dilation of blood vessels, causing a fall in blood pressure.

histology the study of the tissues, tissue structure and organs of living organisms, in the main through microscopic techniques.

histolysis breakdown of a cell or tissue.

HIV (*abbreviation for* human immune deficiency virus) the retrovirus thought to be the cause of AIDS.

hoar frost ice crystals formed on surfaces, e.g. vegetation, cooled by radiative heat loss. The ice comes from frozen dew and the SUBLIMATION of water vapour to ice.

hole *see* **electric current**.

holography a method of recording and reproducing three-dimensional images using light from a LASER, but without the need for cameras or lenses. The holographic images are generated by two beams of laser light producing interference patterns. A single beam of laser, or coherent, light is split into two. One beam is reflected onto the object and then onto the photographic film or plate. The second reference beam passes straight onto the film. The interference pattern on the film produces a **hologram**. The developed film, when illuminated by coherent light, reproduces the image because the interference patterns break up the light, which reconstructs the original object. Because a screen is not required, the light forms a three-dimensional image in air.

homeostasis the various physiological control mechanisms that operate within an organism to maintain the internal environment at a constant state. For example, homeostasis operates to keep the body temperature of humans within a small, crucial temperature range, independent of the temperature of the external environment, as our metabolic processes would not function in any other temperature range.

homologous pertaining to organs or structures that have evolved from a common ancestor, regardless of their present-day function. For example, the pentadactyl limb is the ancestral form of the quadruped forelimb, and from it evolved the human arm, the fin of cetaceans, and the wings of birds. These structures are therefore said to be homologous. Similarities in homologous structures are best seen in early embryonic development and imply relationships between organisms living today.

homologous chromosomes chromosomes that are identical in their genetic LOCI but can have individual allelic forms (*see* ALLELE) that are not necessarily the same. In DIPLOID organisms, a pair of homologous chromosomes exists in all SOMATIC CELLS, each member of the pair having come from a different parent. During MITOSIS in somatic cells, homologous chromosomes do not associate with each other in any way. However, during MEIOSIS in the formation of the GAMETES, homologous chromosomes join together to form a pair, and exchange of genetic material may occur before the homologous pair separates into two new cells that produce gametes. Therefore, the gametes have only a single set of chromosomes so that at fertilization the diploid number is restored.

homologous series chemical compounds that are related by having the same functional group(s) but formulae that differ by a specific group of atoms. For instance, the ALKENES form an homologous series in which each successive member has an additional CH_2 group, i.e.:

Alkene Series	Molecular Formula
Ethene	C_2H_4
Propene	C_3H_6
Butene	C_4H_8

homozygote an organism that has two identical ALLELES of the GENE in question in all SOMATIC CELLS. For instance, if gene B has two allelic forms, B and b, the homozygote will contain only one allelic type, i.e. either BB or bb, at the appropriate location on a pair of HOMOLOGOUS CHROMOSOMES. Homozygotes can thus only produce one kind of gamete, B or b, and as such are capable of pure breeding. For example, the gene for albinism is RECESSIVE, and any individual that possesses this phenotypic trait will be homozygous for the gene. If two such individuals breed, the resultant offspring will all be albinos.

Hooke's law the physical relationship between the magnitude of the applied force on an elastic material and the resulting extension. The extension must be within the YIELD zone of the material, as any force that goes beyond this will cause permanent deformation. Hooke's law can be represented by the equation $T = kx$, where T is the magnitude of force, k is the spring constant, and x is the displacement of the material.

horizon in geology, a general stratigraphic term referring to a plane within a series of rock layers. It has little or no thickness as such and is used to pinpoint lithological changes or may refer to a very thin bed within a unit. In astronomy, the great circle, with the NADIR and ZENITH forming the poles, in which a plane that forms a tangent to the Earth's surface cuts the CELESTIAL SPHERE.

horizontal the term in mathematics that describes a line that is at right-angles with the vertical and parallel to the horizon.

horizontal parallax the value of the GEOCENTRIC PARALLAX when an observer sights a heavenly body on the HORIZON.

hormone an organic substance, secreted by living cells of plants and animals, that acts as a chemical messenger within the organism. Hormones act at specific sites, known as "target organs," regulating their activity and eliciting an appropriate response. In animals, hormones are secreted from various ductless glands, which include the pancreas, thyroid and adrenal glands. This hormone-signalling system is collectively known as the ENDOCRINE SYSTEM. These glands secrete hormones directly into the bloodstream, usually in small amounts, where they circulate until they are picked up by appropriate receptors present on the cell membranes of the target organs. These receptors recognize the particular hormone and bind to it, initiating a response. Hormones also play an important part in the role of plant and seed growth and are found in root tips, buds, and other areas of rapid development. For example, *gibberellins* are a class of plant hormones involved in processes such as initiating responses to light and temperature, the formation of fruit and flowers, and the promotion of seed elongation. Hormone action is constantly regulated by elaborate FEEDBACK MECHANISMS, both within and between cells and organs, that regulate their secretion and breakdown.

Horsehead nebula a well-known DARK NEBULA in the Orion CONSTELLATION.

horst an area of land that is bounded by two or more NORMAL FAULTS with opposite senses of movement, resulting in a central block that is raised, relative to either side.

hot-wire anemometer a device for measuring the speed of wind. It comprises a current-carrying wire that is cooled by the wind, and calculations can be made to relate wind speed to the RESISTANCE of the wire.

hovercraft *see* **friction**.

Hovmüller diagram a diagram representing time on one axis and longitude on the other on which are scales, lines or curves indicating values of an atmospheric variable, e.g. pressure. The diagram shows the movement of large atmospheric features over a period of time.

Hubble constant a measure of the rate of expansion of the universe and its variation with distance, in effect the time period since all matter was together in one mass. Based on work by the American astronomer Edwin Powell Hubble (1889–1953), it is calculated from the RED SHIFTS of distant GALAXIES and produces figures of between 5 and 10 thousand million years.

humidity the amount of water vapour in the Earth's atmosphere. The actual mass of water vapour per unit volume of air is known as the absolute humidity and is usually given in kilograms per cubic metre (kgm^{-3}). However, it is useful to use relative humidity, which is the ratio, as a percentage, of the mass of water vapour per unit volume of air to the mass of water vapour per unit volume of saturated air at the same temperature (*see also* SPECIFIC HUMIDITY).

humus the material in soil that results from the decomposition of animal and vegetable matter. It provides a source of nutrients for plants.

Huntington's chorea *see* **lethal gene**.

hurricane on the BEAUFORT SCALE, a wind with a mean speed of 64 knots or more (74 mph). Also, an intense tropical cyclonic storm in which the winds circulate at high speeds. Hurricanes occur in the Pacific and North Atlantic. In the western Pacific, a hurricane is known as a typhoon.

hybrid rocks "mixed" rocks that are formed through the reaction of MAGMA with the rock (IGNEOUS, METAMORPHIC or SEDIMENTARY) forming the containment for the magma.

hydrides compounds formed by the reaction of elements with hydrogen. A number of compound types are formed, distinguished by bonding and molecular structure. Hydrides of alkali or alkali earth metals are salt-like and crystalline; non-metals often form liquids or gases, many of which dissolve in water to form an acid or alkali; and TRANSITION ELEMENTS form hybrids.

hydrocarbon an organic compound that contains carbon and hydrogen only. There are many different hydrocarbon compounds, the most common being the ALKANES, ALKENES, and ALKYNES.

hydrochloric acid an aqueous solution of hydrogen chloride gas, producing a colourless, fuming, corrosive liquid. It will react with metals to form chlorides, liberating hydrogen. It is made by the ELECTROLYSIS of brine producing hydrogen and chlorine, which are combined, or by the reaction of SULPHURIC ACID with sodium chloride. It has many uses in industry.

hydrogen the lightest element, which forms molecules containing two atoms H_2. It occurs free and is widely distributed as a component of water, minerals and organic matter (*see* HYDROCARBON). It is manufactured by electrolysis of water and is also produced in the catalytic (*see* CATALYST) treatment of petroleum. It is explosive over a wide range of mixtures with oxygen and combines with most elements to form HYDRIDES. It has numerous uses, including HYDROGENATION reactions, organic and inorganic synthesis, e.g. methanol, ammonia and hydrochloric acid, also metallurgy, and filling balloons. It has three isotopes, protium, DEUTERIUM and tritium, containing 0,1 and 2 NEUTRONS respectively.

hydrogenation an important industrial reaction where gaseous HYDROGEN is the vehicle for adding hydrogen to a substance. The reaction usually proceeds in the presence of a CATALYST and often at elevated pressure. This reaction is utilized in the petroleum-refining and petrochemicals industry, the hydrogenation of coal to produce HYDROCARBONS, and the hydrogenation of fats and oils.

hydrogen I and II the hydrogen that has been identified in space and that occurs in two forms.

Hydrogen I (otherwise known as neutral hydrogen) is seen in the spiral arms of the galaxy and emits radio radiations. Hydrogen II (or ionized hydrogen) is found in gaseous nebulae, emitting visible and radio radiations.

hydrogen bomb *see* **nuclear fusion**.

hydrogen peroxide (H_2O_2) a strong oxidizing and bleaching agent usually in the form of a SOLUTION in water. On decomposing it produces water and oxygen, hence it is used as a bleach. It is used industrially and recently it has been used as the oxidizing agent in rocket fuel.

hydrology the study of water and its cycle, dealing with bodies of water and how they change, and all forms of water (rain, snow and surface water), including its use, distribution and properties. It involves aspects of OCEANOGRAPHY, METEOROLOGY and GEOLOGY.

hydrolysis the term used to describe a chemical reaction where the action of water causes the decomposition of another compound and the water itself is decomposed. In salt hydrolysis, the salt dissolves in water, producing a solution that may be neutral, acidic or basic, depending on the relative strengths of the ACID and BASE of the salt. For example, a solution of potassium chloride (KCl) would be neutral, since potassium forms a strong base and chlorine forms a strong acid. In comparison, ammonium chloride (NH_4Cl) gives an acidic solution, since ammonium forms a weak base but chlorine forms a strong acid.

hydrometeor a general term encompassing all forms of water vapour in the atmosphere that have condensed or sublimed, e.g. RAIN, SNOW, FOG, CLOUD and DEW.

hydrometer an instrument, consisting of a bulb with a long stem, that, when floated in a liquid with the stem upright, enables the relative density of the liquid to be measured.

hydrosphere the water that exists on or near to the earth's surface. The main components are water (H_2O), sodium chloride (NaCl) and magnesium chloride ($MgCl_2$). By mass, the major elements are oxygen (almost 86 per cent), hydrogen (10.7 per cent), chlorine (2 per cent) and sodium (1 per cent). Magnesium is the only other element present in significant quantities.

hydrothermal metamorphism the reaction of very hot waters—produced by hydrothermal activity (associated with igneous activity)—with the rocks through which they pass, resulting in the alteration of minerals. Such reactions include serpentinization—the alteration of iron/magnesium minerals in ULTRABASIC IGNEOUS ROCKS to minerals of the serpentine group (hydrated magnesium silicate), and chloritization that also acts upon iron/magnesium minerals in igneous rocks, producing members of the chlorite group (hydrated silicates with magnesium and iron).

hydroxide a compound derived from water (H_2O) through the replacement of one of the hydrogen atoms by another atom or group, e.g. NaOH, sodium hydroxide. ALKALIS are the hydroxides of metals.

hydroxyl the OH group comprising an oxygen and a hydrogen atom bonded together. In alcohols the OH group occurs in a COVALENTly bonded form.

hyetograph an apparatus for the collection and measurement of rainfall.

hygrometer an instrument for measuring the relative HUMIDITY of the air.

hygroscopic the term applied to a substance that absorbs moisture.

hypertonic a term used to describe a liquid that has a higher osmotic pressure (*see* OSMOSIS) than another, or a standard, with which it is being compared.

hypotenuse in a right-angled triangle, this is the name given to the longest side, which always faces the right angle.

hysteresis the effect when a physical process lags behind its cause. For example, when a material or body is stressed, the STRESS produces STRAIN. When the stress is removed, the strain is not removed immediately or completely, and a residual strain remains.

Iapetus Ocean a late Precambrian/early Palaeozoic ocean that lay between what is now northwestern/central Europe on the east and Greenland and North America on the west. During the Caledonian OROGENY, the two continental masses were brought together as the ocean floor was subducted (*see* SUBDUCTION). The tract of deformed rocks (Caledonian orogenic belt) created during the Lower Palaeozoic is found in Norway, Greenland, Scotland, Northern Ireland, Wales, eastern Canada and the eastern U.S.A.

ice age the spreading of glacial ice over areas that are normally ice-free. Most is known about the Pleistocene ice age, but the geological record contains evidence of earlier events although the evidence is less clear.

iceberg a large body of ice, derived from a GLACIER or ice sheet, floating in the sea. Icebergs can be carried long distances before finally melting.

icosahedron a polyhedron with 20 plane faces. If the faces are equilateral triangles, then the icosahedron is said to be regular.

ideal gas a gas that exists only hypothetically and therefore obeys the GAS LAWS precisely. In an ideal gas, the molecules would occupy negligible space, and the forces of attraction between them would also be negligible. The ideal gas would also show perfect elasticity, since it would not be able to store energy, i.e. the molecules would be able to return to their original dimension after any collisions between neighbouring molecules and their containing vessel.

identical twins the offspring that develop from a single fertilized ovum that splits into two very early during development, producing two separate individuals. Identical twins (also known as monozygotic twins) have precisely the same genetic constitution and are always of the same sex.

Ig *abbreviation for* IMMUNOGLOBIN.

igneous rock one of the three main ROCK types. Igneous rocks are formed by the intrusion of magma at depth in various physical forms or extrusion at the surface as lava flows. Typical rocks are GRANITE, BASALT and DOLERITE.

ignimbrite a PYROCLASTIC rock that is poorly sorted and deposited as a hot flow of pumice (frothy volcanic glass) and rock fragments in a fine ash that acts like a fluid. If the temperature is sufficiently high, and there is pressure from numerous flows, the lower layers may become welded to produce a planar FABRIC defined by flattened fragments of pumice. Ignimbrites may be a few hundred metres or many kilometres long and several metres thick.

image there are two optical images—*real* and *virtual*. The images seen in a flat mirror are formed by beams of light diverging from a point behind the mirror, and this sort of image—a virtual image—can never be formed on a screen. However, images created by projection equipment, where the rays of light converge on a screen after passing through a lens, are real images.

imbricate structure a series of rock slices created by and between THRUSTS, making a "stacked-up" arrangement.

immiscible the term that describes liquids that cannot be mixed together. Such liquids tend to be polar and non-polar (e.g. water and ether respectively), which, when added together, form two separate layers, with the less dense liquid forming the upper layer. Conversely, two polar liquids, or two non-polar liquids, will mix together and as such are termed miscible.

immune system the defence system within the bodies of vertebrates, which evolved to afford protection against the pathogenic effects of invading micro-organisms and parasites. The immune system confers two types of immunity to an organism:
(1) Innate (or natural) immunity—this is present from birth and is non-specific, operating against almost any substance that threatens the body.
(2) Acquired immunity—this type of immunity is as a consequence of an encounter with a foreign substance, and it is specific against that foreign substance.

immunoglobin (Ig) these are groups of proteins, collectively termed ANTIBODIES, that are produced by specialized cells of the blood, which are called B-CELLS and which can bind to specific ANTIGENS. B-cells are stimulated to divide in the presence of particular antigens, and the resultant daughter cells produce quantities of immunoglobins, which play an important role in the body's IMMUNE SYSTEM.

impermeable a substance which does not allow gases or liquids to pass through.

incandescence the state when a substance or body

is at a sufficiently high temperature to emit light, e.g. as in the filament of an electric light bulb.

inclusion the condition when a mineral encloses another mineral, gas or liquid. Fluid inclusions contain gas and/or liquid and sometimes a solid phase. The inclusions form during crystallization or recrystallization, and their study can provide information about the origin of the enclosing mineral.

independent assortment a process suggested by the Austrian monk Gregor Mendel to explain the random distribution of different gene pairs that allows all possible combinations to appear in equal frequency. Independent assortment is also known as MENDEL'S second LAW OF GENETICS.

independent variable in a mathematical expression, this term describes a VARIABLE that can take any value, irrespective of how the other quantities are varied. For example, in the equation $y = 3x + 8$, y is a function of x and is thus the DEPENDENT VARIABLE, while x is the independent variable.

index fossil a fossil that is selected for a particular feature and that occurs in a particular ZONE. The zone usually takes the name of the fossil. An index fossil should ideally have a wide distribution in area but not in time. Over the fossil-bearing record, index fossils include TRILOBITES (in the Cambrian) and GRAPTOLITES (Ordovician and Silurian).

index mineral (*see also* BARROVIAN METAMORPHISM) within an area that has been subjected to REGIONAL METAMORPHISM, any of certain typical minerals that appear with increasing grade. A line delineating the first appearance of an index mineral equates to a line of constant metamorphic grade and is called an **isograd**.

indicator a chemical substance, usually a large complex organic molecule, that is used to detect the presence of other chemicals. Indicators are usually weak ACIDS, the un-ionized form (often written as HA), having a different colour than the ionized form (H and A), due to the negative ion A. The degree of ionization, and thus colour change, depends on the pH of the solution under investigation. In solution, the indicator partially dissociates, i.e. HA —> H and A. Most useful indicators give a distinct colour change over a small pH range, usually about two units. Commonly used indicators include phenophthalein and methyl orange.

inductor in chemistry, a substance that reacts rapidly with a reactant in a reaction to facilitate and speed up reaction with the other reactant.

induration hardening through heating, applied especially to sediments intruded by igneous rocks.

inequality the mathematical statement of which of two quantities is the larger or smaller. Thus if x is larger than y, it is written x > y and thus x < y signifies x is smaller than y.

inert gas any of the elements comprising group 8 of the PERIODIC TABLE. The inert gases are helium, neon, argon, krypton, xenon and radon. They are unreactive monatomic gases, which make up about 1 per cent of air by volume, argon being the most abundant. The outer shell comprises eight electrons, which gives these elements their stable configuration since they will not readily lose or gain electrons. Xenon, neon and krypton are extracted from liquid air by FRACTIONAL DISTLLATION, and are used commercially in fluorescent lamps. Helium occurs in natural gas deposits, which is its principal source, and is important in low-temperature research, since it has a boiling point lower than any other substance.

inertia the property of a body that causes it to oppose any change in its present state of motion. Thus, unless a body is acted upon by an external force, it will remain at rest or continue moving at a constant speed in a straight line. The first of NEWTON'S LAWS OF MOTION states that the mass of an object gives a direct measure of its resistance to changing its direction of motion, i.e. its inertia.

infinity (∞) the term used to describe a number or quantity with a value too great to be measured. For example, outer space is regarded as boundless and is therefore described as infinite. Sometimes the symbol is written as -∞, negative infinity, if the value is so small as to be incalculable and insignificant for all practical purposes. Such a value is described as being infinitesimal.

infrared astronomy the study of INFRARED RADIATION (0.8 to 1000μm) emitted from celestial objects. The radiation tends to be absorbed by water vapour in the atmosphere, a problem partially overcome by siting observatories at high altitudes.

infrared detection detection of rays by several methods, including photoconduction of lead sulphide and telluride, photographically with specific dyes, and directly with, for example, THERMISTORS (temperature-sensitive resistance) or THERMOCOUPLES.

infrared radiation electromagnetic radiation with wavelengths between those of visible light and microwaves, i.e. from about 0.75 μm to 1 mm. Such radiation can penetrate fog, and, by using special photographic plates, details invisible to the naked eye may be rendered visible (*see also* ELECTROMAGNETIC WAVES).

inhibitor in chemistry, a substance that will slow or stop a reaction by preventing chain reactions or by rendering surfaces inactive. Similarly, in biology, a substance that reduces the rate of an ENZYME-catalysed reaction, either by binding to the active site of the enzyme or by forming a complex with the substrate/enzyme intermediate.

initiator a substance that starts a chemical reaction. It also applies to highly sensitive explosives used as detonators, e.g. mercury fulminate.

inlier an outcrop of older rock surrounded by younger, which may be formed by faulting or folding and subsequent erosion.

inorganic chemistry the division of chemistry concerned with ELEMENTS and their compounds other than carbon. It thus excludes organic compounds such as alcohols, esters, and hydrocarbons. However, simple carbon compounds, such as metal carbonates, oxides of carbon, and the physical and chemical properties of the element carbon, are usually included.

insolation the Sun's radiation reaching the Earth, which depends, *inter alia*, upon the orbital position of the Earth and the translucency of the atmosphere.

insoluble a term used to describe substances that will not dissolve in a SOLVENT, or that will dissolve to only a very limited extent. As the term is not precise, different parameters may be set, but generally a substance is said to be insoluble if it will dissolve only to the extent of 0.1 gram or less per 100 millilitres of solvent. A slightly soluble substance has a SOLUBILITY of between 0.0 and 1 gram per 100 ml of solvent, whereas a soluble substance has a solubility of 1 gram or more per 100 ml of solvent.

insulator a substance that is good at retaining heat and preventing electric flow since it is a poor CONDUCTOR. This is usually due to the fact that insulators have few mobile ELECTRONS because all the electrons are strongly attracted to the PROTONS in the atoms of the insulator. Materials commonly used as insulators include plastics, ceramics, rubber, and many other non-metals.

insulin a pancreatic HORMONE that initiates glucose uptake by body cells and thus controls glucose levels in the blood. Insulin functions by stimulating certain PROTEINS found on the surface of cells within the vertebrate body to take up glucose, which would otherwise be unable to enter cells, as it is very hydrophobic. *Diabetes mellitus* is a condition in which the blood contains excessively high glucose levels due to an under-production of insulin, and the excess glucose is excreted in the urine. This condition can prove fatal, but sufferers can be successfully treated by insulin therapy.

integer any number that is not a fraction but a whole number. Integers can have positive or negative values or be zero. The following can all be thought of as integers: -4, -3, -2, -1, 0, 1, 2, 3 and 4.

integration a branch of CALCULUS that is used to evaluate an area under the curve of a particular function. There are various methods of integration, ranging from simple formulae, which can be thought of as the inverse of differential formulae, to the more complex formulae that are used to find the inverse of irregular functions.

interference the interaction between two or more waves passing through a medium at the same place at the same time, resulting in specific patterns. Constructive interference occurs between waves that are in phase, i.e. their crests or troughs overlap, producing a wave of greater AMPLITUDE. A wave with maximum amplitude is produced when two waves are precisely in phase. Destructive interference occurs when waves are out of phase, i.e. where a crest and a trough overlap, producing a wave with reduced amplitude. If two waves are exactly out of phase then they cancel each other out. For example, with light waves, interference may produce alternate bright and dark bands—the brightest bands correspond to waves with maximum amplitude (constructive interference) and the darkest bands correspond to waves with minimum amplitude (destructive interference). Similarily with sound, intervals of silence and increased volume correspond to waves displaying destructive and constructive interference respectively. Interference patterns can also be detected in radio waves and are the cause of distortions in reception.

intergalactic medium any material that might exist in space far from any galaxy. It has been suggested that its mass could far outweigh the visible galaxies.

intermediate rocks IGNEOUS ROCKS that compositionally are between acidic and basic rocks, and contain about 53 to 66 percent silica. Typical examples are diorite (coarse-grained rocks with plagioclase FELDSPAR, QUARTZ and ferromagnesian minerals, e.g. PYROXENE or hornblende, an AMPHIBOLE) and andesite (the fine-grained equivalent of diorite).

intermontane basin a sedimentary basin that lies between and is being infilled by sediment eroded from surrounding mountain ranges.

interphase *see* **mitosis**.

interplanetary matter matter in the solar system, such as dust, METEORITES, COMETS but excluding the planets and their satellites. It also includes charged particles from the SOLAR WIND.

interplanetary space space occupied by a "thin" gas and beyond the effects of planets. The gas comprises gas, mainly hydrogen, dust and streams of PROTONS, ELECTRONS and other particles from the Sun.

interstellar medium the matter found throughout space and comprising gas and dust—mainly HYDROGEN, HELIUM and INSTERSTELLAR MOLECULES. In all, it amounts to about 10 per cent of the mass of the GALAXY.

interstellar molecule any of over fifty species of molecule found in the interstellar medium, especially in cold, dense clouds (the congregation of such clouds of gas molecules and dust may lead to the formation of stars). Some of the more common molecules are carbon monoxide (CO), water, ammonia (NH_3), methane (CH_4), methanol (CH_3OH), and formaldehyde (methanal, HCHO).

intertropical convergence zone a narrow zone at low latitude where air masses from north and south of the equator converge, often resulting in depressions that may lead in the ocean areas to HURRICANES when the zone is displaced from the equator.

intrusion a body of (usually) igneous rock that is pushed into existing rocks. An arbitrary division into major and minor is often adopted, the latter including DYKES, SILLS and PLUGS, where one dimension is in the order of metres. Major intrusions (or *plutons*) include BATHOLITHS, igneous DIAPIRS, stocks (roughly circular, steep-sided and discordant, with a domed top) and laccoliths, which have a lensoid shape and push up the strata above them to create a domed roof fed by a vertical pipe from beneath.

inversion *or* **temperature inversion** where the usual temperature gradient in the atmosphere is reversed and temperature increases vertically. This occurs often in anticyclonic conditions, producing stable air near the ground on clear nights.

in vitro a term applied to experiments or techniques undertaken in the laboratory, where biological or biochemical processes are carried out "in glass."

in vivo biological and biochemical processes that occur in a living organism or cell.

ion an ATOM or MOLECULE with a positive charge due to electron loss (a CATION) or a negative charge due to electron gain (an ANION). The process of producing ions is known as ionization, and it can occur in a number of ways, including a molecule dissociating into ions when it is added to a solution, or the formation of ions by bombarding atoms with radiation.

ion exchange the exchange of IONS of like charge between a solution and a highly insoluble solid. The solid (ion-exchanger) consists of an open molecular structure containing active ions that exchange reversibly with other ions in the surrounding solution without any physical changes occurring in the material. For example, ion exchange is also used to soften hard water by removing calcium ions. The water is passed through a column containing an exchange resin that contains sodium ions, and these are exchanged for the calcium ions in the hard water, leaving the water calcium-free.

ionization potential if enough energy is available, it is possible to remove one or more electrons from an atom, molecule or ion (measured in electron-volts). The minimum amount of energy required to remove one electron from a gaseous atom is the ionization potential, and the removal of the second or third electrons are the second or third ionization potentials. The ionization potentials for atoms reflect the binding energies of the outermost electrons and thus are lowest for the alkali metals (e.g. sodium, Na) where the last electron in its structure is in a new quantum shell.

ionization temperature a useful factor in calculating stellar temperatures from spectra. It is the temperature at which the NUCLEUS and ELECTRONS in an atom become detached from each other.

ionosphere the upper ATMOSPHERE (above 80 km/50 miles), which contains a high concentration of IONS and free ELECTRONS, particularly between 100 or 300 km (62 and 186 miles). Radio waves are reflected by certain regions of the ionosphere, thus facilitating transmission around the curvature of the Earth.

iron (Fe) a metallic ELEMENT in group 8 of the PERIODIC TABLE. It occurs naturally as magnetite (Fe_3O_4), haematite (Fe_2O_3), limonite ($FeO(OH)nH_2O$), siderite ($FeCO_3$) and pyrite (FeS_2). It is extracted from its ores by the blast furnace process. Often in combination with other elements, it is the most widely used of all metals.

iron meteorite a meteorite composed of iron and iron-nickel alloy with a very small proportion of silicate and sulphide minerals. Of all meteorites actually seen to fall, this variety is very much in the minority (*see also* STONY METEORITE).

irrational number a quantity that cannot be expressed as a fraction and can only be approximately expressed as a decimal. The square root of 2 ($\sqrt{2}$) is an irrational number, with the approximate decimal value of 1.414. Irrational numbers together with the set of RATIONAL NUMBERS comprise the set of REAL NUMBERS.

island arc a line of volcanoes on the continental side of a deep oceanic trench that marks the SUBDUCTION of OCEANIC CRUST. Almost three-quarters of past or present volcanoes are in a belt around the Pacific, especially along the volcanic island arcs. EARTHQUAKES are associated with these areas, due to the downward movement of the subducted slab, and the gradual meeting of the plate at depth releases MAGMA and fluids that rise to generate the volcanism of the island arcs.

isobar a line used to join points of equal atmospheric pressure on a weather map at a given time. If there is a great change in pressure over a small area then the change in weather is more apparent, and this is shown on a weather map by closely drawn isobars.

isoclinal fold a FOLD that has its limbs parallel.

isograd *see* **index mineral**.

isohel a line joining locations that have the same amount of sunshine.

isohyet a line joining locations that have the same rainfall.

isomer a chemical compound that has the same molecular formula as another chemical compound but differs in the arrangement of the constituent atoms. Isomers are studied in the branch of chemistry known as STEREOCHEMISTRY, which is concerned with the spatial aspects of the structure of molecules. All isomers fall into one of the following two groups:

(1) Structural isomers—isomers that differ in the way their atoms are bonded to each other.

(2) Stereoisomers—isomers that have atoms that are bonded in the same way but differ in the way they are arranged in space. Isomers usually have different chemical properties and different physical properties (melting point, boiling point, density, etc).

isopach a contour line that joins points of equal thickness in a rock unit.

isopach map a geologic map constructed using ISOPACHS to show subsurface geology. The technique may be used to reconstruct an erosional surface or to estimate the morphology of a petroleum reservoir.

isopleth *see* **nomogram**.

isosceles the term that describes a triangle that has two angles and sides of equal magnitude. It also describes a type of TRAPEZIUM, where the two non-parallel sides are equal.

isostasy a term coined by an American geologist, Clarence Edward Dutton (1841–1912), in 1889 whereby the differing heights of the features of the Earth were postulated to have "roots" of differing depths for the purpose of "balance." The proposal is based upon the implication that there is a depth at which the pressure everywhere is the same. To achieve this, high mountain ranges in the CONTINENTAL CRUST have a root projecting into the mantle.

Thinner crust is balanced by a correspondingly greater depth of denser mantle, and the depth of mantle increases further beneath this oceanic crust. In fact, the dynamic movements of PLATE TECTONICS mean that this postulate can rarely be satisfied, but in regions that have been stable for some time, this state of equilibrium is almost reached.

isotherm a line joining locations with the same temperature.

isotonic (solutions) solutions with the same osmotic pressure (*see* OSMOSIS).

isotope an atom that differs from other atoms of the same element due to a different number of NEUTRONS within its nucleus. As isotopes still have the same number of PROTONS, their ATOMIC NUMBER is unchanged, but the varying number of neutrons affects their MASS NUMBER. Most elements exist naturally as a mixture of isotopes but can be separated due to the fact that they have slightly different physical properties. For laboratory purposes, a device called the MASS SPECTROMETER is used to separate the different isotopes.

isotope geochemistry (*see also* GEOCHRONOLOGY) the study of particular isotope ratios in rocks to determine the age of rocks, to study geothermometry (i.e. determining the temperature at which an event took place), and *inter alia* to determine the provenance of ore-forming liquids, MAGMAS and natural waters.

isotropy the feature whereby the properties of a substance do not vary with direction (the opposite of ANISOTROPY).

J

jasper an impure variety of minutely crystalline quartz (SiO_2). It is usually red, brown or yellow, and some varieties are banded e.g. Egyptian or Ribbon Jasper.

jaundice a condition characterized by the unusual presence of bile pigment circulating in the blood. Jaundice is caused by the bile produced in the liver passing into the circulation instead of the intestines because of some form of obstruction. The symptoms of jaundice include a yellowing of the skin and the whites of the eyes.

jet a variety of LIGNITE that occurs as hard, black masses in SHALES rich in organic material. Jet is formed from driftwood deposited in stagnant water and subjected to chemical action. It often shows the structure of wood, and in the past was worked for ornamental pieces, jewellery, etc.

jet stream a high-speed westerly wind in the region of the TROPOPAUSE. The principal jets are the polar front and subtropical jets which occur at heights between 10 and 15 km (6 and 9 miles). The winter jet stream occurs between 50 and 80 km (31 and 50 miles) in the upper STRATOSPHERE. The wind speeds can approach 300 km/h (186 mph).

joint a fracture in rock that may occur as a parallel set or, more commonly, in an irregular and less

203

systematic manner. Where a set of joints can be identified, it can usually be related to tectonic stresses and the geometry of the rock body. There are several types of joint: unloading joints, caused by release of stress on rocks at depth as overlying rocks are removed by erosion; cooling joints, which occur in igneous bodies; and joints related to regional deformation.

joule the unit of all ENERGY measurements. It is the mechanical equivalent of heat, and one joule (J) is equal to a force of one NEWTON moving one metre, i.e. 1J = 1Nm. It is named after James Prescott Joule (1818–1889), a British physicist who investigated the relationship between mechanical, electrical and heat energy, and, from such investigations, proposed the first law of THERMODYNAMICS, the conservation of energy.

Julian calendar the calendar introduced by the Roman emperor Julius Caesar. Based on a year of 365.25 days, it remains the calendar in use today, although somewhat modified.

Julian date a system for the consecutive numbering of days, irrespective of month and year, which is used especially in astronomy. The starting point was midday on 1 January, 4713 BC, and the system was introduced by the French scholar Joseph Justus Scaliger (1540–1609) in 1582. (There is no connection with the Julian calendar).

Jupiter the largest planet in the SOLAR SYSTEM and fifth from the SUN (*see* APPENDIX 6 for physical data) with its orbit between MARS and SATURN. Its atmosphere is made up of HYDROGEN with approximately 15 percent HELIUM and traces of water, ammonia and methane, which becomes a liquid shell surrounding a zone of metallic hydrogen (that is, the hydrogen is compressed so much that it behaves like a conducting metal) around a rock and ice core that is ten times the mass of Earth.

A longstanding feature of the planet is the Great Red Spot, which has been observed since the 17th century and is thought to be a storm. There are 16 satellites, the largest of which are Io, Europa, Ganymede and Callisto (the Galilean satellites). These are visible with an ordinary telescope.

juvenile water water that originates from a MAGMA and that has never been in the atmosphere. Surprisingly, water in great quantities can originate in this way.

K

kame a structure produced by glacial deposition (*see* GLACIAL DEPOSITS). It occurs as a hummock of sands and gravels, with bedding and often exhibiting slumping at the sides. It was formed by the melting of stagnant ice, which caused the load to be dropped.

kaolin *see* **china clay** and **kaolinite**.

kaolinite a hydrated aluminium silicate that occurs as an alteration product in IGNEOUS rocks, GNEISSES and a weathering and hydrothermal reaction product in sedimentary rocks. It is used in ceramics and as a coating and filler in the paper industry.

kaolinization the high temperature hydrothermal alteration and breakdown of FELDSPARS to form KAOLINITE. The process can proceed in granites to the point where the rock literally falls apart and the only recognizable, original mineral is quartz (*see* CHINA CLAY).

karst a karst landscape is one created in a limestone area and created by the limestone itself. The distinctive landforms are a result of dissolution by solutions that move through joints and fissures and that dissolve calcareous material, enlarging cracks and joints and creating underground waterways and caves. Features typical of a karst include networks of furrows and sharp crests, funnel-shaped hollows, and ultimately conical hills and steep-sided depressions of impressive scale.

karyotype the number, shapes and sizes of the CHROMOSOMES within the cells of an organism. Every organism has a karyotype that is characteristic of its own species, but different species have very different karyotypes. For example, all normal human females have 22 pairs of DIPLOID chromosomes with similar shape and size, but all female horses have 32 pairs of diploid chromosomes with their own unique shape and size.

katabatic wind the sinking and downward movement of cold, dense air beneath warmer, lighter air. The air is cooled by radiation, usually at night. It occurs over ice-covered surfaces and in the fjords of Norway, and in many cases can be gale force.

kelp a large, brown, algal seaweed found anchored to the sea bed below low-tide level. Kelp is a source of iodine and potash.

Kelvin scale the unit of temperature (K) based on the temperature scale devised by the British physicist Lord William Kelvin (1824-1907). The Kelvin scale has positive values only, with the lowest possible unit of 0K, which is equal to –273.15°C or –459.67°F.

Kepler's laws of planetary motion a set of laws, named after the German astronomer Johannes Kepler (1571–1630) and published early in the 17th century, that state:

(1) The planets move about the Sun in ellipses with the Sun at one of the foci of the ellipse.

(2) A line joining the Sun to each planet covers equal areas in equal times.

(3) The square of the planet's year and the cube of

its average distance of the Sun are the same proportion for all planets.

keratin a fibrous, sulphur-rich protein consisting of coiled POLYPEPTIDE chains, which occurs in hair, hooves, horn and feathers.

kerogen a fossil organic material, bituminous in nature, that is found in oil shales. As with all oil shales, the HYDROCARBONS are produced by destructive DISTILLATION of the shales.

kerosene *or* **kerosine** a thin oil that is one of many products obtained during the FRACTIONAL DISTILLATION of PETROLEUM. Kerosene is used as fuel for jet engine aircraft.

ketenes ORGANIC compounds with the group $>C=C=O$ of which the first member is ketene, $CH_2 = CO$. They are very unstable and polymerize easily. They react with water to produce ACIDS, and with ALCOHOLS to give ESTERS.

ketone an organic compound that contains a $C=O$ (carbonyl) group within the compound, as opposed to either end of the compound. There are many forms of ketones, and their physical and chemical properties differ due to the presence of alkyl groups ($-CH_3$) or aryl groups ($-C_6H_5$) within the ketone molecule. Ketones can be detected within the bodies of human beings when fat stores are metabolized to provide energy if food intake is insufficient. If these accumulate within the blood, the undernourished person will experience headaches and nausea. The presence of ketones in urine is called **ketonuria**.

kettle hole a hole or depression formed in glacial drift due to outwash material from a GLACIER covering isolated masses of ice. When the covered ice eventually melts, the sediments slump down into the space, creating a surface depression.

kilobyte (KB) *see* **byte**.

kilocalorie a unit of heat used to express the energy value of food. One kilocalorie is the heat needed to raise the temperature of one kilogram of water by 1°C. It is estimated that the average person needs 3000 kilocalories per day, but this requirement will vary with the age, height, weight, sex and activity of the individual.

kilogram a unit of mass (kg) that is equal to the international prototype made of platinum and iridium stored in the French town of Sèvres.

kimberlite a highly fragmented ULTRABASIC IGNEOUS rock formed from a MAGMA of relatively rare composition. The surface occurrence of kimberlite is in pipe-like diatremes, which are volcanic vents formed explosively. The mode of formation and the deep source mean that material from considerable depths is often incorporated into kimberlites and pieces of the wall rock are broken off as the magma rises. This renders kimberlites very interesting petrologically as the XENOLITHS from depth give clues as to the composition of the MANTLE. Many xenoliths contain minerals formed at high pressure, e.g. diamond.

kinesis the response of an organism to a particular stimulus in which the response is proportional to the intensity of the stimulation.

kinetic energy the energy possessed by a moving body by virtue of its mass (m) and velocity (v). The kinetic energy (Ek) of any moving body can be determined using the following equation:

$$Ek = \frac{1}{2}mv^2$$

(the energy is in joules if m is kg and v ms^{-1})

209

As the kinetic theory of matter states that all matter consists of moving particles, it holds that all particles must possess some amount of kinetic energy, which will increase or decrease with the surrounding temperature (*see* EVAPORATION).

Klinefelter's syndrome a condition in which human males have the abnormal GENOTYPE of XXY rather than normal XY. This produces recognizable characteristics within the affected male, such as the development of breasts and smaller testes (resulting in reduced fertility). Klinefelter's syndrome occurs in approximately one in a thousand male births and is caused by nondisjunction of sex chromosomes during MEIOSIS.

klippe a structure analogous to an OUTLIER but formed initially due to tectonic movements and occurring as a detached part of a THRUST sheet, and bounded by thrusts.

kneejerk reflex a complex neural pathway in humans in which a blow just beneath the kneecap (the patella) results in a rapid extension of the leg. The knee jerk is a response by the central nervous system to stimulation of sensory NEURONS located at the front of the thigh. The response leads to the contraction of one muscle (quadriceps, at the front of the thigh) and the inhibition of the contraction of another muscle (biceps, at the back). As the quadriceps muscle is the extensor, the leg straightens but is prevented from bending as the biceps muscle, which is the flexor, is inhibited from contracting.

knot a unit of nautical speed equal to one NAUTICAL MILE (1.15 statute miles or 1.85 kilometres) per hour. The term knot originates from the period

when sailors calculated their speed by using a rope
with equally spaced tied knots, attached to a heavy
log trailing behind the ship. The regular space
between knots was measured at 47 feet, 3 inches,
which is 14.4 metres.

Köppen classification a system of climatic classification that was developed between 1910 and 1936,
and is based upon annual and monthly means of
temperature and precipitation, and the major types
of vegetation. The system comprises three orders or
levels, beginning with the overall climate, e.g. warm,
temperate and rainy (class C), which can then be
categorized further, as for example—winter, dry
produces class Cw. The third level qualifies temperature, for example a hot summer would produce
a classification Cwa.

The major climates are: tropical rainy (A), dry (B),
warm temperate rainy (C), cold snowy (D), and
polar (E). (*See also* THORNTHWAITE CLIMATE CLASSIFI-
CATION).

Korsakoff's syndrome *or* **Korsakoff's psychosis**
a neurological disorder first described by a Russian
neuropsychiatrist called Sergei Korsakoff (1854-
1900). The condition is characterized by gross defects in memory for recent events, disorientation,
and no appreciation of time. Patients with
Korsakoff's syndrome are unaware that there is a
problem and are liable to confabulate. Although it
can result from lack of vitamins or lead and manganese poisoning, Korsakoff's syndrome most commonly occurs as a complication of chronic alcoholism. It is caused by a dietary deficiency of vitamin
B1 (thiamine) which is needed for the conversion of
carbohydrate to glucose. The syndrome has been

invaluable in neuropsychology, as it has helped to
identify the brain regions involved in the memory
processes of recall and recognition.

Krebs cycle *see* **citric acid cycle**.

L

labelled compound a compound used in radioactive tracing, where an atom of the compound is replaced by a radioactive ISOTOPE, which can be followed through a biological or physical system by means of the RADIATION it emits (*see also* TRACER).

labile likely to change, not stable. A term used in chemistry.

Lagrangian point the point at which a particle of little mass can remain fixed between two objects orbiting a common centre of gravity.

Lagrangian points in particular, with respect to the Earth and Moon, the locations where the gravitational forces are equal and so objects positioned at such points remain fixed. There are five such points between the Earth and Moon.

laevorotatory compound any substance that, when in crystal or solution form, has an optically active property that rotates the plane of polarized light to the left (anticlockwise).

lahar a mudflow developed on the flank of a volcano under the combined effects of eruption and torrential rainstorms (or melting of ice or snow). The majority of volcanic fatalities are due to lahar, which can travel many kilometres from the source. If an eruption interacts with a crater lake producing a hot mudflow, then the result can be even more catastrophic.

213

lamination rock stratification where the units measure less than one centimetre, and are commonly of the order of a few millimetres.

land and sea breezes air circulation along coasts during summer, developed when the overall pressure gradient is minimal. During the day the Sun warms the land more than the adjacent sea, and so the air above the land becomes warmer and thus rises. This produces a convective motion, with the cooler air from the sea moving onto the land. At night the situation is reversed because the sea is warmer than the land.

Langmuir's theory the basic premise, essentially correct, that electrons in an atom are arranged in shells that correspond to the periods of the PERIODIC TABLE, and that the most stable configuration is a complete shell.

lanthanides (*otherwise known as the* **rare earth elements**) these elements, from lanthanum (La) to lutetium (Lu), have much in common, chemically, with the scandium group (group 3B of the PERIODIC TABLE). The properties of these metals are very similar, and the lanthanides and yttrium (symbol Y) all occur together and are separated by CHROMATOGRAPHY. The elements are reactive with the heavier ones resembling calcium, while scandium is similar to ALUMINIUM.

lapilli (*singular* **lapillus**) small fragments of lava ejected from a volcano and ranging in size from 2 to 64 mm (0.08–2.5 inches). Lapilli may be magmatic, e.g. PUMICE, or they may contain rock fragments. Accretionary lapilli are layered pellets of ash that formed in volcanic ash clouds before transportation.

laser (*acronym for* Light Amplification by Stimulated Emission of Radiation) a device that produces a powerful and narrow monochromatic beam of light. Lasers are used extensively in electronic engineering, fibre optic communications, HOLOGRAPHY, and in certain surgical operations.

Lassaigne's test a chemical test for the presence of nitrogen and also SULPHUR or HALOGENS. The sample is heated with sodium in a test tube, quenched and ground, and on reaction with certain reagents a characteristic product or colour is produced.

latent heat the measurement of heat ENERGY involved when a substance changes state. While the change of state is occurring, the gas, liquid or solid will remain at constant temperature, independent of the quantity of heat applied to the substance (an increase in heat will just speed up the process). The specific latent heat of fusion is the heat needed to change one kilogram of a solid into its liquid state at the MELTING POINT for that solid. For example, the specific latent heat of fusion for pure, frozen water (ice) at 273K (0°C) is 334 kJkg⁻¹. The specific latent heat of vaporization is the heat needed to change one kilogram of the pure liquid to vapour at its boiling point. In the case of pure water again, at its boiling point of 373K (100°C), 2260kJkg⁻¹ is the specific latent heat of vaporization needed to change water into steam.

lateral line the sensory system of fish and aquatic amphibians, which comprises sensory cells (neuromast organs) in a line along the body. Movements in the water affect the neuromasts, thus creating nerve impulses.

lateral moraine rock debris created by a GLACIER, which accumulates at the margin of a valley glacier and is due to transport and reworking of rocks from the valley sides. It eventually accumulates as MO-RAINE.

latex originally used to describe the fluid obtained from rubber trees, which is essentially a SUSPENSION of rubber in water. It is also applied to synthetic polymers and to rubbers that are produced as latexes and that may be used for the manufacture of goods.

latitude the angular distance of a particular point on the earth's surface relative to the earth's equator. Latitude is measured in degrees corresponding to the angle of incident light from a specific star, e.g. the Sun, above the horizon at a given time and is described as being north or south of the equator. On a world globe, lines of latitude are represented by parallel, horizontal lines.

lattice a specific structural arrangement of atoms in a CRYSTAL. The smallest complete lattice, which is repeated throughout the crystal, is called the unit cell.

lattice energy the energy needed to break down one MOLE of a substance from its crystal LATTICE into the gaseous state where the constituent ions are very far apart.

Laurasia one of two continental masses produced by the rifting of PANGAEA. Laurasia was the northern area, which became North America, Greenland, Europe and Asia, and the splitting apart is thought to have occurred in the early Mesozoic (*see* APPENDIX 5), about 240 million years ago.

Laurent series a mathematical expression that is particularly useful in the analysis of an area between two CONCENTRIC circles. The given function is expressed as both a positive and a negative infinite POWER SERIES, using the following formula:

$$f(z) = \sum_{n=-\infty}^{\infty} a_n(z-a)^n$$

lava molten rock erupted by a VOLCANO, whether on the ground (subaerial) or on the sea floor (submarine). Both acidic and basic forms are found and show a number of textures, but all are characterized by some glass and/or fine-grained minerals due to rapid cooling. The way it is erupted and moves, and therefore its subsequent morphology, depends greatly on the VISCOSITY; generally, a less viscous lava will flow faster. Two types of basaltic lava forms are seen: "aa," which is jagged, and "pahoehoe," which exhibits a smoother, ropy appearance.

law of conservation *see* **energy, thermodynamics**.

LD50 the amount of a toxic substance, which, when applied in a specific manner, will kill 50 per cent of a large number of individuals within a species.

LDL *see* **lipoprotein**.

LDL-receptor *see* **cholesterol**.

Le Chatelier's principle a statement relevant to chemical reactions, which predicts that if the conditions of a system in EQUILIBRIUM are changed, the system will attempt to reduce the enforced change by shifting equilibrium.

lee wave a stationary (STANDING) wave in air in the lee of a mountain barrier, created by air passing over the mountain and then returning to its original level. Sometimes such waves may have large AMPLITUDES, and clouds often form along the wave crests.

legionnaires' disease an infectious disease caused by the bacterium *Legionella pneumophila*, which inhabits surface soil and water. It has also been traced in water used in air-conditioning cooling towers. The main source of infection is inhalation of air or water carrying the bacteria, and so far there is no evidence that it is transmitted from an infected to a non-infected individual. Legionnaires' disease is really a form of pneumonia, and thus its symptoms include shortness of breath, coughing, shivering and a rise in body temperature. Healthy individuals should fully recover from infection if treated with the antibiotic called erythromycin.

Leibniz's theorem a mathematical formula that states that the nth derivative $[uv]^n$ of the product of the function u and the function v can be calculated using the following:

$$\sum_{i=0}^{n} \binom{r}{i} u^{(i)} v^{(n-1)}$$

This allows the limit of the first n terms to be reached. The derivative of the product of two functions, u and v, is obtained using the PRODUCT RULE. Gottfried Wilhelm Leibniz (1646–1716) was a philosopher and scientist born in the German town of Leipzig. He devised differential and integral CALCULUS and developed certain aspects of DYNAMICS. He contributed to the development of many other mathematical, philosophical and geological theories, such as the BINARY number system and the theory that present-day earth evolved from a molten origin.

lens a device that makes a beam of rays passing through it converge or diverge. In optics, a lens is a uniform transparent refracting medium shaped with two curved surfaces, classified according to the nature of these surfaces. Each surface can be convex, concave or plane. The optical axis is a line joining the centres of curvature of the lens surfaces (which are the centres of the spheres of which the lens surfaces form a part), and the optical centre is a point on the axis in the lens which does not deviate a ray passing through.

Lenz's law the principle devised by the German physicist Heinrich Friedrich Emil Lenz (1804–65) to deduce the direction of an induced current during ELECTROMAGNETIC INDUCTION. Lenz's law states that an induced current will flow in a direction that will oppose the change that induced the current. Lenz's law explains the phenomenon of a suspended closed coil of wire always repelling an approaching magnet but attracting a departing magnet, independent of the orientation of the magnet.

Leonids a meteor swarm whose ORBIT crosses the Earth's orbit.

leprosy an infectious disease that affects the skin, nerves and mucous membranes of the patient. The symptoms of leprosy include severe lesions of the skin and destruction of nerves, which can lead to disfigurements such as wrist-drop and claw-foot. Leprosy is caused by the airborne bacterium, *Mycobacterium lepra*, and, fortunately, is not highly contagious as transmission involves direct contact with this bacterium. The likeliest source of infection, therefore, arises from the nasal secretions (swarming with bacteria) of patients and not from the popular misconception of touching the skin of an infected individual. Leprosy is curable, and the treatment, using sulphone drugs, has the beneficial side-effect of making the patient non-infectious even if he or she is not completely cured. Although the incidence of leprosy was once worldwide, it is now mostly confined to tropical and subtropical regions.

lepton *see* **elementary particle**.

lethal gene a gene that, if expressed, will cause the death of the individual. The fatal effect of the expressed gene usually occurs in the prenatal developmental stage of the individual, i.e. the embryonic stage for animals and the pupal stage for insects. Although most examples of lethal mutants fail to survive to adulthood, there is one well-researched genetic disorder, called Huntington's chorea, which does not usually affect the individual until middle age. Huntington's chorea is caused by a single dominant gene, and thus half the children of an affected parent will inherit the genetic disorder, although fortunately this disease is rare.

leucocratic a descriptive term used to specify a light colour in igneous rocks due to the dominance of QUARTZ or FELDSPAR in the main over the darker minerals, e.g. PYROXENES, AMPHIBOLES and BIOTITE.

leucocyte a large, colourless cell formed in the bone marrow and subsequently found in the blood of all normal vertebrates. It is commonly known as a white blood cell and plays an important role in the IMMUNE SYSTEM of an individual. Leucocytes are produced in the bone marrow, spleen, thymus and lymph nodes of the body and can be classified into the following three groups in order of decreasing constitutency of leucocytes:

Group	%	Functions
Granulocyte	70	Helps combat bacterial and viral infection and may also be involved in allergies.

Group	%	Functions
LYMPHOCYTE	25	Destroys any foreign bodies either directly (T-CELLS) or indirectly by producing antibodies (B-CELLS).
Monocyte	5	Ingests bacteria and foreign bodies by the mechanism called PHAGOCYTOSIS.

leucoplast a colourless object that contains starch and is found in some plant cells. If a leucoplast contains the pigment CHLOROPHYLL it may develop into a CHLOROPLAST.

leukaemia a cancerous disease in which there is an uncontrolled proliferation of white blood cells (LEUCOCYTES) in the bone marrow. The white blood cells fail to mature to adult cells and thus they cannot function as an important part of the defence system against infections. Although the definite cause of leukaemia is as yet unknown, there is growing suspicion that certain viruses may cause it and that perhaps there is an hereditary component. Unfortunately, leukaemia is not a curable disease, but there are methods effective in suppressing the reproduction of white blood cells—radiotherapy and, more commonly, chemotherapy. These methods bring the disease under control and thus help prolong the patient's life.

levée a bank or ridge running along the edges of a river and sloping away from the water. It is caused by deposition of coarse sands and silts when the river floods. If repeated, it may permit the river level to rise above its flood plain with the risk of greater damage should further flooding occur.

Lewis acid in a chemical reaction, any substance accepting an ELECTRON PAIR.

Lewis base in a chemical reaction, any substance donating an ELECTRON PAIR.

librations the oscillation of a satellite, e.g. the Moon, when viewed from Earth. It is due to a PARALLAX effect.

lichen a FUNGUS and an ALGA in a symbiotic (*see* SYMBIOSIS) association that has produced a form different from both. The cells of the alga are protected within the plant body created by the fungus and photosynthesize, passing food to the fungus. Lichens grow slowly but can tolerate extreme climates, and in arctic regions provide a valuable food for reindeer. Most species are also sensitive to air pollution (especially sulphur dioxide) and have been used as indicators.

lidar a device that uses a LASER beam to detect or observe distant cloud patterns by the amount of BACKSCATTER from the beam.

ligand any MOLECULE or ATOM capable of forming a bond with another molecule (usually a metallic CATION) by donating an ELECTRON PAIR to form a complex ION. In biological terms, ligand refers to any molecule capable of binding with a specific ANTIBODY.

light curve the graphical plot obtained from measuring the change in brightness of a VARIABLE STAR with time.

lightning the discharge of high-voltage electricity between a cloud and its base, and between the base of the cloud and the earth. One flash of lightning actually consists of several separate strokes that follow each other at intervals of fractions of a second (too fast for the human eye to detect), and occurs when the strength of the ELECTRIC FIELDS becomes great enough to overcome the RESISTANCE of the intervening air. The clap of thunder that can be heard during thunderstorms is a result of the expansion of the intervening air.

light reactions the biochemical processes that generate ATP, oxygen and a reduced coenzyme called NADPH during PHOTOSYNTHESIS in the presence of light. The light-dependent reactions occur in the inner membranes of CHLOROPLASTS and require water and several forms of the pigment CHLOROPHYLL. Two of the products of the light reactions, ATP and NADPH, enter the CALVIN CYCLE, whereas the third product, OXYGEN, is released by the plant.

light wave *see* **electromagnetic waves**.

light year a measure of the distance travelled by light in one year, which is approximately 9.467×10^{12} kilometres.

lignite *or* **brown coal** a coal that in RANK falls between peat and bituminous coal. It shows little alteration, the woody structure often being visible. It has a high moisture content, which can be reduced, and is found in Tertiary basins in Britain, and in Europe.

limb in geology, the area between hinges in folded rocks (*see* FOLD) In astronomy, the rim of a heavenly body that has a visible disk, e.g. the Sun or Moon.

limestone a SEDIMENTARY rock that is composed mainly of CALCITE with DOLOMITE. Limestone may be organic, chemical or detrital in origin. There is a tremendous variety in the make-up of limestones, which may comprise remains of marine organisms (corals, shells, etc); minute organic remains (as in CHALK); or grains formed as layered pellets (*see* OOLITE) in shallow marine waters and also calcareous muds. Recent deposits of CALCIUM CARBONATE are found in shallow tropical seas. Modification after deposition is usually extensive, and both the composition and structure may change due to compaction, recrystallization and replacement.

Limestone has many uses commercially, including building stone, roadstone and aggregate, as a source material in the chemicals industry, and in its natural state as AQUIFERS and petroleum RESERVOIR rocks.

limiting reactant any substance that limits the quantity of the product obtained during a chemical reaction. The limiting reactant can be identified using the chemical equation of a reaction as it will be the smallest quantity in comparison to the other reactants and products.

limonite a secondary "mineral," approximating in composition to $FeO.OH_nH_2O$. However, it is now known to be a rock composed of goethite ($Fe_2O_3.2H_2O$) and lepidocrocite ($FeO.OH$). It is usually amorphous and, as a weathering product of mineral deposits or iron in rocks, can occur in considerable deposits. It is the major constituent of bog iron ores, which form on lake floors due to bacterial action.

lineament a long structural or volcanic feature on the Earth's surface.

linear equation a mathematical term used to describe any equation containing VARIABLES that are not raised to any power although the variables may have a COEFFICIENT. For example, the following equations are linear combinations of a,b,c and x,y,z respectively:

$$a + 2b = 4c \text{ and } 2x + 3y + z = 0.$$

lineation (the linear equivalent of FOLIATION) a set of linear structures produced in a rock by deformation or surface processes. There are numerous types of lineation, and these may be due to: striations caused by movement along a surface, e.g. SLICKENSIDES; the axes of small FOLDS or crenulations; elongation of deformed objects such as pebbles; orientation of minerals, especially elongate varieties: or intersection of planar features (intersection lineations).

line of sight velocity the velocity at which a heavenly body approaches or recedes from the Earth.

line squall when warm air overruns cold air it produces a squall line, often associated with a cold front, along which there are stormy conditions. It is characterized by sudden changes in temperature and pressure producing wind, heavy cloud and thunderstorms. The squalls may occur simultaneously, in a line across-country.

linkage the association between two or more GENES situated on the same CHROMOSOME. The genes tend to be inherited together, thus the parental gametes, which after FERTILIZATION eventually form the offspring, will not have undergone normal GENETIC RECOMBINATION to generate new combinations of genes. As the distance separating two genes decreases, the chance of these two genes becoming separated during CROSSING-OVER also decreases, thus increasing the chance that these genes are linked to be inherited as one segment rather than separate genes.

lipase any enzyme capable of breaking down fat to form FATTY ACIDS and GLYCEROL. Lipases function in alkaline conditions and are most abundant in the pancreatic secretions during digestion.

lipids an all-embracing term for oils, fats, waxes and related products in living tissues. They are ESTERS of FATTY ACIDS and form three groups: simple lipids, including fats, oils and waxes; compound lipids, which includes PHOSPHOLIPIDS; and derived lipids which includes STEROIDS.

lipoprotein (LDL) any protein that has a FATTY ACID as a side chain. Lipoproteins have significant importance in certain biological processes as they function as a transport mechanism for essential molecules. For example, as CHOLESTEROL is extremely hydrophobic, there would be no method of transporting it to its target body tissues. This problem is solved by low-density lipoprotein surrounding the cholesterol molecule and forming a hydrophilic molecule that can be transported by body fluids.

liquefaction of gases a gas may be turned into its liquid form by cooling below its critical temperature (the temperature above which a gas cannot be liquefied by PRESSURE alone). In addition, pressure may be required. For gases such as oxygen, helium and nitrogen, low temperatures are used.

liquid a fluid state of matter that has no definite shape and will acquire the shape of its containing vessel as it has little resistance to external forces. A liquid can be regarded as having more KINETIC ENERGY than a SOLID but less kinetic energy than a GAS. It is considered that the average kinetic energy will increase as the temperature of the liquid rises.

liquid crystal one of certain liquids that show regions of aligned molecules that are similar to crystals. The application of a current disrupts the molecules sufficiently to darken the liquid and form, for example, characters on a display.

lithification the processes that change unconsolidated sediment into rock, including cementation of the grains.

lithology the description of a rock through its grain size, structure, mineral content (as far as possible) and general appearance.

lithosphere that layer of the Earth, above the ASTHENOSPHERE, which includes the crust and the top part of the mantle down to 80–120 kilometres (50–75 miles) in the oceans and around 150 kilometres (93 miles) in the continents. The base is gradational and varies in position depending upon the tectonic and volcanic activity of the region. The lithosphere comprises the blocks that constitute PLATE TECTONICS.

litmus a natural compound obtained from lichens, which is used as an INDICATOR, turning red to indicate acid conditions and blue for alkaline.

litre a unit of volume given the symbol l and equal to 1000 cubic centimetres, i.e. $1l = 1000cm^3 = 1dm^3$. One gallon is approximately 4.5 litres.

loam a type of soil with sand, silt and clay in roughly equal proportions, often with some organic matter.

local gravitational constant the quantity given to the ACCELERATION of any object near to sea level at any point on earth. This acceleration is a result of gravity and is given the symbol g. It is calculated using the following:

$$g = \gamma MR^{-2}$$

where $g = 8MR^{-2}$ = universal gravitational constant, M = mass of the earth, R = radius of the earth at that point.

At the North Pole, for example, $g = 9.8321 \text{ ms}^{-2}$, and at the equator $g = 9.7801 \text{ ms}^{-2}$.

local group the cluster of galaxies that includes the Milky Way. There are approximately 20–25 members within 5 million light years.

locus (*plural* **loci**) the set of specific points that either satisfies or is determined by a certain mathematical condition. The locus can be thought of as tracing the path of a moving point relative to another fixed point. For example, a circle is a locus of a point that moves in such a way that the distance (radius of circle) between the moving point and the fixed point (centre of the circle) is constant. In biology, the locus is the name given to the region of a CHROMOSOME occupied by a particular GENE.

lode a vein or fissure in a rock containing mineral deposits.

loess a sediment formed from the aeolian transportation and deposition (*see* AEOLIAN DEPOSITS) of mainly silt-sized particles of QUARTZ. It is well sorted, unstratified and highly porous, and although it can maintain steep to vertical slopes, is readily reworked. Loess is widespread geographically, and although thicknesses of Chinese deposits exceed 300 metres (984 ft), it is normally a few metres. Its origin has been hotly debated, but it is now accepted that the loess particles are produced by glacial grinding, frost cracking, hydration in desert regions and aeolian impact of sand grains. The wind is the primary and essential agent in the process.

logarithm (*abbreviation* **log**) a mathematical FUNCTION that was first introduced as a labour-saving device when dealing with the multiplication and division of large numbers. However, the modern calculator has reduced this need for logarithms, and now a logarithm is the power of a number to a specified base written in the form $\log_a x = n$ (a=base). There are two forms of logarithm:

(1) Common logarithm—has base 10, $\log_{10} x$

(2) Natural or Napierian logarithm—has base e, lnx (where e is EXPONENTIAL FUNCTION).

lone pair a pair of electrons which are not shared by another atom but which can form CO-ORDINATION COMPOUNDS.

longitude the angular distance of a given point on the Earth's surface relative to the Greenwich meridian (a theoretical line that runs through the North and South Poles as well as Greenwich, England). Longitude is measured in degrees east and west of the Greenwich meridian (this is given an arbitrary value of 0°). By international agreement, the world is divided into 24 longitudinal zones, each 15° in width, starting and finishing at Greenwich (0°), to be able to relate the different times in places throughout the world. Thus, any place in the zone centred at 15° east of Greenwich is one hour ahead of Greenwich, whereas any place in the zone centred at 15° west is one hour behind Greenwich.

longitudinal wave the classification for a wave that is produced when the vibrations occur in the same direction as the direction of travel for that wave. The most well-known example of a longitudinal wave is a SOUND wave, which is propagated by displacement of air particles, causing areas of high density (called compressions) and areas of low density (called rarefactions). The WAVELENGTH of a longitudinal wave is the combined length of a single compression and rarefaction. For the FREQUENCY and speed of a longitudinal wave, *see* WAVE.

longshore drift the movement of material (sand and shingle) along the shore by a current parallel to the shore line (i.e. a longshore current). Longshore drift occurs in two zones: beach drift, at the upper limit of wave activity (i.e. where a wave breaks on the beach); and the breaker zone (where waves collapse in shallow water) where material in suspension is carried by currents.

Lotka's equations mathematical expressions used in ecology for the relationships between two species competing for food or space in the same area, and the situation when one is predator and one prey.

loudspeaker a TRANSDUCER that converts electric current into sounds through the vibration of a paper cone or diaphragm.

low an area of low pressure—*see* DEPRESSION.

lubricant any substance that, when applied between surfaces, will cause a decrease in FRICTION. Some common lubricants include oil and graphite dust.

luminescence the emission of light by a living organism that is not a consequence of raising the body temperature of the organism. For luminescence to occur, the cells of the organism must contain the protein called luciferin, the enzyme luciferase, and the energy source for the reaction, ATP.

luminosity the amount of light radiated by a star, expressed as a magnitude.

lymph a colourless, watery fluid that surrounds the body cells of vertebrates. It circulates in the LYMPHATIC SYSTEM but is moved by the action of muscles, as opposed to the contraction of the heart. Lymph consists of 95 per cent water but contains protein, sugar, salts, and LEUCOCYTES, and transports fat from the gut wall during digestion.

lymphatic system a network of tubules found in all parts of a vertebrate's body except, if present, the central nervous system. The tubules drain the body fluid, LYMPH, from the tissue spaces, and as they gradually unite to form larger vessels, the lymph vessels finally drain into two major lymphatic vessels that empty into veins at the base of the neck. The system consists of vessels and nodes, where the fluid is filtered, bacteria and other foreign bodies are destroyed, and LYMPHOCYTES enter the lymphatic system. The lymphatic system is not only an essential part of the IMMUNE SYSTEM, but also carries excess protein and water to the blood and transports digested fats.

lymphocyte a cell that is produced in the bone marrow and is an essential component of the IMMUNE SYSTEM as it will differentiate into either a B-CELL or a T-CELL. Lymphocytes are a form of white blood cell (LEUCOCYTE), which collect in the lymph nodes of the lymphatic system and defend the body against foreign bodies and bacterial infections. Lymphocytes are also present in blood but in a smaller percentage.

lysis the destruction of cells, commonly BLOOD CELLS, by antibodies called **lysins**. More generally, the destruction of cells or tissues by pathological processes, e.g. autolysis, where cells are broken down by enzymes produced in the cells undergoing breakdown.

lysozyme an enzyme that is present in tears, nasal secretions and on the skin, and has an antibacterial action. It breaks the cell wall of the bacteria, leaving them open to destruction. Lysozyme also occurs in egg-white, and it was the first enzyme for which the three-dimensional structure was determined.

M

Mach number the speed of a body expressed as a ratio with the SPEED OF SOUND. Mach 1 is sonic speed, and thus subsonic and supersonic are below and above unity, respectively.

mackerel sky a cloud pattern comprising wavy CIRRO- or ALTOCUMULUS with holes, suggesting the markings of a mackerel.

Maclaurin's formula a mathematical formula devised by Colin Maclaurin, an 18th-century Scottish polymath, used to expand certain functions around zero. When a function f(x) is infinitely differentiable and has real values, then it can be estimated as the sum of f(0) and the first few terms of the series using the following:

$$f(x) = f(o) + xf'(o) + \frac{x^2}{2!} f''(o) + \cdots + \frac{x^r}{r!} f^{(r)}(o) + \cdots$$

macrophage a specialized cell that forms part of the IMMUNE SYSTEM of vertebrates. Macrophages are derived from MONOCYTES in the blood system and can move to infected or inflamed areas of the body using PSEUDOPODIA. In such areas, they ingest and degrade broken cells and other debris including microbes by means of PHAGOCYTOSIS.

Magellanic clouds two separate GALAXIES, detached from the Milky Way, which appear, from the southern hemisphere, as patches of light. They are approximately 180,000 LIGHT YEARS away and contain a few thousand million stars.

magma the fluid rock beneath the earth's surface, which solidifies to form IGNEOUS ROCKS. During volcanic eruptions, the lava extruded at the earth's surface is not necessarily the same in composition as the magma that arises to form lava, since the magma may have lost some of its gaseous elements and some solids of the magma may have crystallized.

magmatic differentiation *see* **differentiation**.

magnetic anomaly the magnetic field remaining after taking away the effect of the Earth's magnetic DIPOLE, due to concentrations or deficiencies of magnetic rocks and minerals in the crust.

magnetic bubble a portion of computer memory that consists of a small region in a material such as GARNET (a silicate mineral), which is magnetized in one direction. Slices of this material placed on a SUBSTRATE produce a magnetic CHIP that under a magnetic field produces magnetic bubbles. Information can be stored on the chip, which may contain up to one million bubbles in 20mm^2, in BINARY form, through the presence or absence of a bubble in a specific location.

magnetic field the region of space in which a magnetic body exerts its force. Magnetic fields are produced by moving charged particles and represent a force with a definite direction. There is a magnetic field covering all of the earth's surface, which is believed to be a result of the iron-nickel core.

magnetic storm a disturbance of the Earth's magnetic field due to solar particles, after a solar FLARE.

magnetic stratigraphy *or* **magnetostratigraphy**
a branch of stratigraphy based on reversal of the
Earth's magnetic field with time. The mechanism of
SEA-FLOOR SPREADING has been corroborated by magnetic stratigraphy. It was found that the magnetic
field over the ocean's floor (*see* MID-OCEANIC RIDGE)
showed a striped magnetic pattern parallel to the
ridge with rapid changes between stripes. It was
discovered that the changes in magnetization between stripes corresponded with reversals of the
Earth's polarity, which could be dated by comparison with continental rocks. The stripes thus act as
a magnetic stratigraphy, dating much of the oceanic
crust, and in addition they provide evidence for
PLATE TECTONIC motion at both constructive and
destructive margins.

magnetosphere the space around the Earth, and
other planets with a magnetic field, in which charged
particles are affected by the magnetic field of that
planet rather than of the Sun. The extent of the
space is greater on the side of the planet away from
the Sun by a factor of about 4. The boundary of the
magnetosphere is well defined but is altered by
solar activity.

magnetostratigraphy *see* **magnetic stratigraphy**.

magnitude the absolute value or length of a physical
or mathematical quantity.

In astronomy, a relative measure of the apparent
brightness of stars. Based upon a logarithmic scale,
a difference of one in magnitude is actually a ratio
of 2.51, thus a star of this magnitude is 2.51 times
brighter than a star of a lower magnitude. The
absolute magnitude is the magnitude a star would
have if it were placed 10 PARSECS away from Earth.

main sequence a band within the HERTZSPRUNG-RUSSELL DIAGRAM in which most stars lie. The stars range from high temperature and LUMINOSITY to low temperature and luminosity (depending mainly upon mass). During much of its life, while converting HYDROGEN to HELIUM, a star lies in the main sequence and then moves out as the hydrogen is consumed, becoming a RED GIANT initially before evolving into other forms (*see* STELLAR EVOLUTION).

malaria an infectious disease caused by certain parasites in the blood of the victim. Malaria is transmitted by an infected mosquito biting a human, thus injecting the parasite from the salivary gland of the mosquito into the bloodstream of the human. After the parasites have developed in the victim's liver, they are released into the bloodstream and attack red blood cells. Early symptoms of malaria include headaches, body aches and chills. As the disease progresses, malarial attacks are frequent and cause sickness, dizziness and sometimes delirium in the victim; the attacks seem to coincide with the bursting of infected red blood cells. Fortunately, there are many highly effective drugs for treating malaria.

malleability the property of metals and alloys that enables them to be changed in shape by hammering or rolling (or similar processes) into thin sheets.

manganese nodules small (on average 2 cm/0.79 inch) nodules containing manganese oxides, iron oxides with nickel, copper, zinc, cobalt, and other minor elemental traces. The nodules, found on sea and lake floors, are particularly associated with the Pacific Ocean. The nodules show concentric layering, and were probably formed by the alteration of organic matter and other substances in the sediment of the sea floor. The concentration of valuable metals means the nodules are a source of ore, but their recovery is far from straightforward as they are found at depths of up to 4500 metres (14,764 ft) and across vast areas.

manometer an instrument to measure pressure, often in the form of a U-tube containing mercury.

mantle that layer of the Earth between the crust and core. The boundary is marked by the MOHOROVICIC DISCONTINUITY, and the mantle reaches thicknesses of about 2500 kilometres . Its composition approximates to that of garnet peridotite (peridotite is an ULTRABASIC IGNEOUS rock with the minerals olivine, pyroxene and chromite, $FeCr_2O_4$).

map unit the region between two gene pairs on the same CHROMOSOME, where one per cent of the possible products during MEIOSIS are recombinant.

marble a metamorphosed, recrystallized LIMESTONE that does not have a FOLIATION. The alteration renders marble hard enough to take a polish, and it has been and still is used extensively for building, ornamental and decorative work.

maria (*singular* **mare**) the term, from Latin, for the "seas" on the surface of the MOON. The name was coined before modern study found them to be dry, but their origin is not established although it is thought they date from 3300 million years ago. The tendency has been to dispense with the Latin, e.g. Sea of Showers instead of Mare Imbrium.

marker horizon *or* **bed** a thin bed that is easily distinguished, and identified over a large area, making it useful as a stratigraphic correlator. The beds may represent specific depositional events, e.g. a thin coal seam or a volcanic TUFF.

Mars (*see* APPENDIX 6 for physical data) the fourth planet in the solar system with its orbit between Earth and Jupiter. It has a very small atmosphere (pressure less than one hundredth of the Earth's) of CO_2 and frozen CO_2 in polar ice caps. The northern crust is basaltic (*see* BASALT), and there are many extinct volcanoes, canyons and impact craters *and* evidence of water EROSION at some time in the past. It has two small satellites, Phobos and Deimos.

maser (*acronym for* Microwave Amplification by Stimulated Emission of Radiation) a microwave amplifier/oscillator working in a similar way to the laser. Maser oscillations produce a narrow beam of monochromatic (very narrow frequency band) radiation.

mass the measure of the quantity of matter that a substance possesses. Mass is measured in grams (g) or kilograms (kg).

mass number (A) the total number of PROTONS and NEUTRONS in the nucleus of any atom. The mass number therefore approximates to the RELATIVE ATOMIC MASS of an atom.

mass spectrometer the machine used to detect the various types of ISOTOPES found in an element. The mass spectrometer bombards molecules with high-energy electrons, creating smaller positive IONS and neutral fragments. These positive ions are then deflected, using a MAGNETIC FIELD that separates them according to their MASS. The ions finally pass through a slit to the ion collector, and each peak of the printed chart corresponds to a particular ion and its mass. The mass spectrometer also provides information on the relative amount of each isotope present as well as the exact mass of the various isotopes.

mast cell a large, blood-borne cell that has a fast-acting role in the IMMUNE SYSTEM. In allergies, mast cells will be triggered to release histamine when the IMMUNOGLOBIN, IgE, has already attached itself to the foreign body that has entered the body.

matrix (*plural* **matrices**) an array of elements set out in rows and columns. Matrices are a mathematical tool useful in, for example, solving the transformation of co-ordinates. The order of a matrix refers to the number of rows and columns, e.g.:

$$A = (4\ 1\ 6) \quad B = \begin{pmatrix} 2 & 6 \\ 1 & 4 \end{pmatrix}$$

Matrix A has one row, three columns, and matrix B has two rows, two columns. Only matrices of the same order can be added or subtracted. Matrices are only compatible for multiplication if the number of columns of the first equals the number of rows of the second. To multiply, therefore, the row of one matrix is multiplied by the column of the other matrix, and the products are added.

matter any substance that occupies space and has MASS: the material of which the universe is made.

maturity a geological term, referring to a stage in the development of topographic relief, and introduced by the American geographer William Morris Davis (1850–1934) early in the 20th century and now largely superseded. Maturity was the part of the "cycle of erosion," when most of the initial landscape had been eroded away to leave gentle slopes and river valleys.

Me *see* **methyl**.

mean daily motion the angle a celestial body moves through in one day, assuming uniform orbital motion.

meander the side-to-side wandering of a stream or river channel, best developed in river deposits on the FLOOD PLAIN. The origin of meanders is not established but involves the original stream course, the natural hydrodynamic properties of water flowing over sediment or rock, and the fact that the river takes a course that requires the least energy to follow. Once established, meanders create erosion and deposition on the outer and inner banks respectively, and the curved river form can become more exaggerated and "migrate" downstream. A variety of features may be developed, including OXBOW LAKES.

mean solar day the average value of the interval between successive returns of the SUN to the MERIDIAN.

megabyte (MB) *see* **byte**.

megaparsec a unit for defining distance of objects outside the galaxy, equal to 10^6 PARSECS (or 3.26×10^6 light years).

meiosis a chromosomal division that produces the germ cells (GAMETES in animals and some plants, sexual spores in fungi). Meiosis involves the same stages as MITOSIS, but each stage occurs twice, and, as a consequence, four HAPLOID cells are produced from one DIPLOID cell. Meiosis is an extremely important aspect of sexual reproduction, as the production of haploid cells ensures that during FERTILIZATION the chromosomal number is constant for every generation. It also gives rise to genetic variation in the daughter haploid cells by the rearrangement and GENETIC RECOMBINATION if genetic variability already exists in the parent diploid cell.

melanin a dark pigment responsible for colouring the skin and hair of many animals, including humans. Differences in skin colour are due to variations in the distribution of melanin in the skin and not to differences in the number of cells, **melanocytes**, that produce the pigment.

melanism when some individuals in a population are black due to overproduction of the pigment MELANIN. It is particularly well noted with insects found in some industrial areas where, due to the increase in industrial pollution, a species evolved to compensate. In a case of *industrial melanism*, the peppered moth was found to be much darker in polluted areas compared to specimens in non-polluted parts, thus making it less visible to a predator.

melanocratic (*see* LEUCOCRATIC) a descriptive term used to specify a dark colour in IGNEOUS ROCKS due to the dominance of darker minerals (PYROXENES, BIOTITE, AMPHIBOLES, etc.) over lighter minerals such as quartz and feldspar.

melting point the temperature at which a substance is in a state of equilibrium between the solid and liquid states, e.g. ice/water. At 1 atmosphere pressure, the melting point of a pure substance is a constant and is the same as the freezing point of that substance. At constant pressure, the melting point of a substance is lowered if it contains impurities (the reason for adding salt to make ice melt). Although an increase in pressure lowers the melting point of ice, in most substances increased pressure will raise their melting points.

membrane potential the difference in electric potential between the inside and the outside of the plasma membrane of all animal cells. Membrane potentials exist due to the different ionic concentrations within and outwith the cell, and also the selective permeability of the plasma membrane to specific ions (notably potassium [K^+], sodium [Na^+] and chloride [Cl^-] ions). When measured using a microelectrode, the resting potential of most muscle and nerve cells is -60mV on the inside of the plasma membrane.

Mendel's laws of genetics laws of heredity deduced by Gregor Mendel (1822–1884), an Austrian monk, who discovered 16 basic rules experiments with generations of pea plants. Mendel discovered that a trait, such as flower colour or plant height, had two factors (hereditary units) and that these factors do not blend but can be either dominant or recessive. Without knowledge of genes or cell division, he developed the following laws of particular inheritance:

First law—each factor segregates from each other into the GAMETES. It is now recognized that ALLELES are present on HOMOLOGOUS CHROMOSOMES, which separate during MEIOSIS.

Second law—factors for different traits undergo INDEPENDENT ASSORTMENT.

During gamete formation, the segregation of one gene pair is independent of any other gene pair unless the genes are linked (*see* LINKAGE).

meniscus the effect of surface tension on a liquid where it meets a solid, producing a rising of the water's surface up the tube. The reverse effect is seen with mercury.

Mercury (*see* APPENDIX 6 for physical data) the first planet of the SOLAR SYSTEM, and nearest the SUN. It has no atmosphere and a surface temperature of about 425°C. It probably has a relatively large core, producing a high metallic iron : silicate ratio.

mercury a silvery-white liquid metal element that occurs as the ore cinnabar (HgS). As an alloy with most metals, it forms AMALGAMS. It is used extensively in thermometers, barometers and other scientific apparatus, and in the manufacture of batteries, drugs, and chemicals including pesticides, etc. Mercury and many of its compounds are toxic.

mercury barometer an instrument for measuring atmospheric pressure denoted by a column of mercury (in a tube) that exerts an equal pressure.

meridian the great circle that cuts the CELESTIAL SPHERE at its poles and cuts the observer's horizon at the north and south points.

meson an unstable particle belonging to the group called HADRONS. Mesons are found in COSMIC RAYS and consist of a QUARK and its antiquark. Their mass is between that of an electron and a nucleon (a PROTON or NEUTRON), and there are positive, negative and neutral varieties.

mesophyll the internal tissues of a leaf that are between the upper and lower epidermal layers. Mesophyll tissue contains CHLOROPLASTS, which are concentrated at a site that allows maximum absorption of light for PHOTOSYNTHESIS.

messenger RNA (mRNA) a single-stranded RNA molecule that contains ribose sugar, phosphate and the following four bases—ADENINE, CYTOSINE, GUANINE, and URACIL. Messenger RNA has the important role of copying genetic information from DNA in the nucleus and carrying this information in the form of a sequence of bases to RIBOSOMES in the cell, where TRANSLATION occurs to synthesize the specified protein that was encoded in the nuclear DNA.

Messier catalogue a list, compiled by the French comet hunter Charles Messier (1730–1817) in 1770, which contains 108 galaxies, nebulae and star clusters. The names assigned to objects are frequently still used in astronomy.

metabolism the chemical and physical processes occurring in living organisms, which comprise two parts: CATABOLISM and ANABOLISM. ENZYMES control metabolic reactions, which tend to be similar throughout the plant and animal kingdoms.

metal a substance that has a "metallic" lustre or sheen and is generally ductile, malleable, dense and a good CONDUCTOR of electricity and heat. Elements with these properties are generally electropositive, i.e. give up electrons, becoming positively charged (e.g. Na^+) when combining with other RADICALS. When combining with water, BASES result (e.g. NaOH, sodium hydroxide), and their chlorides (e.g. NaCl, sodium chloride) are stable towards water. Not all elements normally considered metals show all these properties. Elements with characteristics of both metals and non-metals are termed METALLOIDS.

metal fatigue *see* **fatigue of metals**.

metalloid an ELEMENT exhibiting some properties associated with metals and some associated with non-metals. Metalloids also exhibit AMPHOTERISM.

metallurgy the scientific study of metals and their alloys, including extraction from their ORES, and processing for use.

metamorphic rock one of the three main ROCK types, formed by the alteration or recrystallization of existing rocks through the primary agents of temperature and/or pressure. There are four types, depending upon the pressure/temperature regime in operation: regional, which are formed by the high pressure and temperature accompanying orogenic (mountain-building) events; contact metamorphic rocks, which are generated by proximity to an IGNEOUS intrusion with high temperature but low pressure; dynamic, which are generated during faulting and thrusting (the sliding of rock masses against each other, producing intense pressure); and burial metamorphism, which is due to high pressures with low temperatures.

metamorphosis the period of change in form of an organism from the larval to the adult state.

metaphase a stage in MITOSIS or MEIOSIS in cells of EUCARYOTES. During metaphase, the chromosomes are organized and attached to the equator of the spindle by their CENTROMERES. Metaphase occurs only once in mitosis but twice in meiosis.

metasomatism the introduction of chemical constituents in a gaseous or liquid phase into a rock (or their removal from it) thus altering the bulk compositon. Many occur with metamorphic processes, and certain minerals may be completely transformed.

metastable a term applied to a mineral assemblage created under high temperature and/or pressure, or a supersaturated solution (*see* SUPERSATURATION) that appears to be stable but, in fact, will react or change if disturbed. The state is due to the system being very slow in reaching equilibrium. A PHASE is termed metastable if it co-exists with another phase that is stable. There are many minerals and whole rock assemblages, e.g. diamond, and a high pressure and temperature METAMORPHIC rock respectively, that

are metastable at the surface. Such assemblages
are temporarily in equilibrium.

metastasis the process of malignant cancerous cells
spreading from the affected tissue to create secondary areas of growth in other tissues of the body.

meteor *or* **shooting star** a small body from the
SOLAR SYSTEM that burns up on entering the Earth's
atmosphere.

meteoric water water originating in the atmosphere, e.g. as rain and snow. (*see also* JUVENILE
WATER).

meteorite *see* **asteroid**.

meteorology the study of the processes and conditions (e.g. pressure, wind speed, temperature) in
the earth's ATMOSPHERE. The resulting data enables
predictions to be made as to likely future weather
patterns.

meteor stream meteor showers caused by dust
streams intersecting the Earth and its atmosphere.
The streams of dust orbit the Sun.

methane the first member of the HOMOLOGOUS series
of ALKANES. It is a colourless, odourless gas with the
chemical formula CH_4. Methane is the main constituent of coal gas and is a byproduct of any decaying vegetable matter. It is flammable and is used in
industry as a source of hydrogen.

methyl *or* **Me** the group CH_3.

methylation the process of adding a METHYL group to
a compound. In biology, it is the addition of a methyl
group to a NUCLEIC ACID base, e.g. ADENINE or CYTOSINE.

metrology the science of measurement.

mica a group of aluminium, potassium SILICATES with
magnesium and iron in dark varieties, and sodium,
lithium or titanium in others. HYDROXYL is always

present, often partly replaced by fluorine. Micas have a sheet-like structure and a perfect cleavage parallel to the sheets (which are made up of Si_4O_{10} groups). The group comprises several members including BIOTITE, muscovite or white mica (potassium variety) lepidolite (lithium potassium), phlogopite (magnesium), zinnwaldite (lithium) and glauconite (iron and extra aluminium). Micas, especially biotite and muscovite occur in many rock types, especially igneous and metamorphic rocks. Muscovite and phlogopite are important in the electrical industry as insulators.

micelle a term used to describe a submicroscopic aggregate of molecules in a colloidal solution (*see* COLLOID). It is applicable especially to soaps and dyes.

micrometer any instrument used for the accurate measurement of minute objects, distances or angles.

micrometre the unit of length that equals 10^{-6} of a metre and has the symbol μm.

microtubule a long, hollow fibre of PROTEIN that is found in all higher plant and animal cells. Microtubules have different functions in different cells, e.g. they form the spindle during MITOSIS, give strength and rigidity to the tentacles of some unicellular organisms, and they are also found in parts of nerve cells.

microwave the part of the electromagnetic spectrum (*see* ELECTROMAGNETIC WAVES) with a wavelength range of approximately 10^{-3} to 10m and a frequency range of 10^{11} to 10^7Hz. When absorbed, microwaves produce large amounts of heat, a useful property for the economical and quick cooking of food. Microwaves are easily deflected, and as they have a shorter wavelength range than radiowaves they are more suitable for use in RADAR systems as they can detect smaller objects.

microwave background a weak radio signal, discovered in 1963, that is thought to be the remains of the big-bang with which the UNIVERSE began. It occurs throughout space at an almost identical intensity.

mid-oceanic ridge, oceanic ridge, *or* simply **ridge** throughout the world's oceans are long, linear volcanic ridges, in effect submarine mountain chains, generally positioned centrally. The ridges are sites where new oceanic crust is created through the spreading of the plates (a spreading axis) and outpouring of BASALT. The mid-oceanic ridges are sites of shallow earthquakes, and they are often cut by TRANSFORM FAULTS that create an offset appearance to the ridge. Those faults linking the ridge are also the focus of earthquakes (*see* SEA-FLOOR SPREADING).

migmatite a rock found in medium to high grade metamorphic regions, which has prompted considerable debate as to its origins. It is a heterogeneous, coarse-grained, mixed rock with a texture resembling a GNEISS, often consisting of a flow-banded rock showing pale-coloured quartzofeldspathic bands (leucosomes) with darker bands of mafic (i.e. iron/magnesium) minerals (constituting the melanosomes). It is now generally considered that most migmatites originate through partial melting at very high temperatures, and they may thus represent the initial stages in MAGMA generation.

Milankovitch theory the theory that periodic changes in the amount of solar radiation received by the Earth led to a cycle of changes in climate and led to ice ages. It was postulated that the variations in radiation were due to, amongst other things, the eccentricity of the Earth's orbit.

Milky Way *see* **galaxy**.

mimicry an adaptively evolved resemblance of one species to another species. Mimicry occurs in both the animal and plant kingdoms but is predominantly found in insects. The main types of mimicry are:

(1) Batesian mimicry, named after the British naturalist H. W. Bates (1825–92)—where one harmless species mimics the appearance of another, usually poisonous, species. A good example of this is the non-poisonous viceroy butterfly mimicking the orange and black colour of the poisonous monarch butterfly. The mimic benefits as, although harmless, any predator learns to avoid it as well as the poisonous species.

(2) Mullerian mimicry, named after the German zoologist J. F. T. Müller (1821–97)—where different species, which are either poisonous or just distasteful to the predator, have evolved to resemble each other. This resemblance ensures that the predator avoids any similar-looking species.

mineral a substance, usually inorganic, with a definite and characteristic chemical composition and typically with a crystalline structure and certain physical properties including: hardness (*see* MINERALOGY), lustre (i.e. the way its surface reflects light), CLEAVAGE, colour, FRACTURE, and relative density. Some elements occur as minerals in their own right, e.g. gold, diamond (carbon), but the vast majority of minerals are compounds of several elements, with oxygen being the commonest element, present as oxides in many cases. There are approximately two thousand minerals, but the most common rocks contain individual minerals such as quartz, calcite, and minerals from about five or six groups, including the clay minerals, PYROXENES, AMPHIBOLES, FELDSPARS and the MICAS. These are all silicate minerals, which form the most abundant of the rock forming minerals.

mineralogy the study of any chemical element or compound extracted from the earth. Mineralogy examines the mode of formation and physico-chemical properties of minerals. These are generally solid or crystalline and can be classified according to their chemical constitution, i.e. molecular, metallic or ionic composition or crystallography. Another method of classifying minerals is according to their comparative hardness, and the Mohs' scale arranges them in relative order from the softest, talc (no. 1) to diamond (no. 10). Each mineral will scratch a mineral lower on the scale. The full list is: 1—talc; 2—gypsum; 3—calcite; 4—fluorite; 5—apatite; 6—orthoclase; 7—quartz; 8—topaz; 9—corundum; and 10—diamond.

mirage a visual phenomenon due to the REFLECTION and REFRACTION of light. A mirage is seen wherever there is calm air with varying temperatures near the earth's surface. A common mirage in the desert is caused by the refraction of a downward light ray from the sky, so that it seems to come from the sand and, to any onlooker, it would appear that the sky is reflected in a pool of water. As the inverted image of a distant tree is also usually formed, the overall effect resembles a tree being reflected in the surrounding water.

mist water droplets in suspension, which reduce visibility to not less than 1 km (1093 yards) (*see also* FOG).

mitochondrion (*plural* **mitochondria**) a double-membrane bound ORGANELLE found in EUCARYOTIC cells. Mitochondria contain circular DNA, RIBOSOMES and numerous ENZYMES, which are specific for essential biochemical processes. Thus, the mitochondrion is the site of such processes as haem synthesis (forms part of HAEMOGLOBIN) and the stages of AEROBIC RESPIRATION that generate most of a cell's ATP. Due to their ATP-producing function, mitochondria are especially abundant in very active, and hence high energy-requiring, cells such as muscle cells or the tails of human sperm cells.

mitosis the process by which a NUCLEUS divides to produce two identical daughter nuclei with the same number of CHROMOSOMES as the parent nucleus. Mitosis occurs in several phases:

(1) Prophase—the condensed chromosomes become visible, and it is apparent that each chromosome consists of two CHROMATIDS joined by a CENTROMERE.

(2) METAPHASE—the nuclear membrane disappears, a spindle forms, and each chromosome becomes attached by its centromere to the equator of the spindle fibres.

(3) Anaphase—each centromere splits, and one chromatid from each pair moves to opposite poles of the spindle.

(4) TELOPHASE—a nuclear membrane forms around each of the group of chromatids (now regarded as chromosomes), and the cytoplasm divides to produce two daughter cells. The stage before and after mitosis is called interphase, and the chromosomes

are invisible during this phase as they have decondensed. This is an extremely important part of a cell's cycle as DNA replicates and required proteins are synthesized during interphase.

mobile belt *or* **orogenic belt** a term applied particularly to older rock sequences (back to the mid-Precambrian) that show stable cratonic (*see* CRATON) zones and unstable mobile belts. The belts comprise highly deformed sediments, many and varied igneous rocks, and slices of crustal material moved up into the belts. The CONTINENTAL crust of present stable areas is made up of worn-down orogenic belt from past episodes of tectonism.

Mobius strip a ribbon of paper where one end has been twisted through 180 degrees before joining to the other end. The result is a single continuous surface containing a continuous curve.

moderator a substance used in a nuclear reactor to slow down fast NEUTRONS generated by NUCLEAR FISSION. The substances contain a light element e.g. DEUTERIUM in heavy water, which absorbs some energy upon impact with the neutron, but avoids capturing the particle. These slower neutrons are then more likely to participate in the ongoing fission process.

modulus the measure of the MAGNITUDE of a quantity regardless of its sign. The modulus of a REAL NUMBER $|x|$ always gives a positive value, e.g. $|-6| = 6$. The modulus of a complex number $|a + ib|$ is the square root of the sum of the squares of the real part (a) and the imaginary part (ib, where $i = \sqrt{-1}$). Thus for the complex number, $|6 + i8|$, the modulus is as follows:

$$|6 + i8| = \sqrt{6^2 + 8^2} = \sqrt{36 + 64} = \sqrt{100} = 10$$

Mohorovičić discontinuity (*often called simply the* **Moho**) a major seismic discontinuity discovered in 1909 from study of the seismograms of the Yugoslav earthquake of that year. The feature was highlighted by double sets of the two types of seismic waves, interpreted as the direct and refracted versions, the first caused by the crust and the second by the mantle. Although this "boundary" was at first considered to be quite a distinct feature, it is now known to be more complex.

Mohs' scale *see* **mineralogy**.

molality the concentration of a solution expressed in MOLES of solute per one kilogram of solvent. Molality has symbol m and units of mol kg^{-1}.

molarity the number of MOLES of a substance dissolved in one litre of solution. Molarity has symbol c and units mol litres^{-1}.

molasse sediments deposited in mainly shallow marine or non-marine conditions and originating from the erosion of mountain belts in the process of being uplifted. The shallow marine sediment would be deposited on beaches, in deltas and sand-bars, while the non-marine phase would produce alluvial fans (*see* ALLUVIUM) and lake deposits in the centre of the depositional basin.

mole the amount of substance that contains the same number of ELEMENTARY PARTICLES as there are atoms in 12 grams of carbon. The number of particles in one mole is called AVOGADRO'S CONSTANT and equals 6.023 x 10^{23} mol^{-1}.

molecular biology the study of the structure and function of large molecules in living cells and the biology of cells and organisms in molecular terms. This applies especially to the structure of PROTEINS and the nucleic acids DNA and RNA.

molecular electronics the use of molecular materials in electronics and opto-electronics. Presently, liquid crystalline materials (liquid crystals) are used in displays, organic semiconductors in xerography, and other organic materials are used in devices where the macroscopic properties of the materials are involved. The long-term aim of this mixed discipline is to devise molecules or small aggregates of molecules that can exceed the performance of silicon-integrated circuits.

molecular formula the chemical formula that indicates both the number and type of any atom present in a molecular substance. For example, the molecular formula for the alcohol ETHANOL is C_2H_6O and indicates that the molecule consists of two carbon atoms, six hydrogen atoms and one oxygen atom.

molecularity the number of particles that are involved in a single step of a reaction mechanism. One step of a mechanism may be termed unimolecular, bimolecular or termolecular, depending on whether there are one, two or three reacting particles. The reacting particles can be ATOMS, IONS or MOLECULES.

molecular sieve a compound with a framework structure, usually a synthetic ZEOLITE, that can be used for the specific absorption of water, gases, etc. In effect, the framework creates a cage to trap the appropriate molecules, and the size of the trap can be modified.

molecular weight the total of the atomic weights of all the atoms present in a molecule.

molecule the smallest chemical unit of an element or compound that can exist independently. Any molecule consists of ATOMS bonded together in a fixed ratio, e.g. an oxygen molecule (O_2) has two bonded oxygen atoms and a carbon dioxide molecule (CO_2) has two oxygen atoms bonded to one carbon atom.

mole fraction the ratio of the number of MOLES of a substance to the total moles present in a mixture.

moment the moment of a FORCE (or VECTOR) about a point is the force multiplied by the perpendicular distance between the point and the line of action. Thus, if a force acts on a point within a body, the action of the force will cause the body to slide and twist around. The tendency of the force to produce a twist about any given point is measured by the moment, *about that point*.

momentum the property of an object that is directly proportional to both its MASS (m) and VELOCITY (v). Momentum (p) is a VECTOR quantity calculated using the equation $p = mv$ (units $kgms^{-1}$ or Ns^{-1}). Changes in momentum mainly occur as the result of an interaction between two bodies, and during any interaction the momentum of a body is always conserved, provided no external force such as FRICTION acts on it. If a body is in rotational motion around an axis, then it is said to have angular momentum. Angular momentum of an object is the product of its momentum and its perpendicular distance from the fixed axis.

monadnock *see* **peneplain**.

monobasic acid an ACID that contains only one replaceable hydrogen atom per molecule. A monobasic acid will produce a normal SALT during a reaction with a suitable metal.

monocline a special type of FOLD, which is asymmetric with one LIMB very short in relation to the others. It is normally applied to large-scale folds.

monoclonal antibody a particular ANTIBODY produced by a cell or cells derived from a single parent cell, i.e. a clone (with each cell being monoclonal). Antibodies thus produced are identical and have specific AMINO ACID sequences. Monoclonal antibodies are used in identifying particular ANTIGENs within a mixture, for example, in identifying blood groups. They are also used to produce highly specific vaccines.

monocotyledon the subclass of flowering plants that have a single seed leaf (cotyledon). Any flowering plant that contains its seeds within an ovary (sometimes forming a fruit) belongs to one of two subclasses, either monocotyledon or dicotyledon (which has two seed leaves). Monocotyledons also differ from dicotyledons in that they have narrow, parallel-veined leaves, their vascular bundles are scattered throughout the stem, their root system is usually fibrous, and their flower parts are arranged in threes or multiples of threes. Most monocotyledons are small plants such as tulips, grasses or lilies.

monocyte a large phagocytic cell (*see* PHAGOCYTOSIS), capable of motion, which is present in blood. Monocytes originate in the bone marrow and, after a short residence in the blood, move into the tissues to become MACROPHAGES.

monomer a simple molecule that is the basic unit of POLYMERS. Most monomers contain carbon-carbon double bonds but just have single bonds after they have undergone ADDITION POLYMERIZATION to form the long-chained polymer. ETHENE is a type of monomer that reacts with other ethene molecules under high pressure and temperatures to form the polymer, polyethene (polythene).

monosaccharides (*see also* SACCHARIDES) sugars with the general formula $C_nH_{2n}O_n$ (where n = 5 or 6). They are grouped into hexoses ($C_6H_{12}O_6$) or pentoses ($C_5H_{10}O_5$) on the number of CARBON atoms present, and are either aldoses, which contain the CHO group, or ketoses containing the CO group.

monsoon in general terms, winds that blow in opposite directions during different seasons of the year, with features that are associated with widespread temperature changes over land and water in the subtropics. Monsoon winds are essentially similar in origin to LAND AND SEA BREEZES, but occur on a very much larger scale geographically and temporally. The Indian subcontinent is subjected to a rainy season in the southwesterly monsoon. Other areas affected by monsoons are Asia (east and southeast), parts of the West African coast and northern Australia.

monozygotic twins *see* **identical twins**.

Moon the Earth's one satellite, at an average distance from Earth of 384,000 kilometres (238,600 miles). It has no atmosphere, water or magnetic field, and surface temperatures reach extremes of 127°C (261°F) and -173°C (-279°F). The surface is heavily cratered, probably due to meteorite impact, and a distinctive feature of the Moon is its MARIA. Rock collected on Apollo space missions has indicated an age of about 4000 million years. A feldspathic crust up to 120 kilometres (75 miles) in thickness overlies a mantle of silicates, and basaltic (*see* BASALT) LAVAS cover almost one fifth of the surface. There is probably a small core of iron, with a radius of approximately 300 kilometres (186 miles).

moraine a general term for ridges of rock debris that

were deposited by GLACIERS and mark present or former ice margins, although it originally referred to ridges of debris around alpine glaciers. There are numerous forms of moraine, from LATERAL MORAINE to terminal moraine (deposited at the leading edge of an active glacier). A medial moraine is a merging of lateral moraines due to the convergence of two glaciers, and ground moraine includes a wavy surface of TILL, BOULDER CLAY or glacial DRIFT. There are essentially two types of moraine when considering formation. Dump moraine is where material is literally dumped (during glacier retreat) at the margin or end of a glacier when the ice is stationary. Push moraines denote the edges of ice advance when the ice moves over sediment and pushes up ridges. Large moraines of this type can be seen, often formed annually, and stratified sediments may be faulted and folded due to the force of the glacier.

mordant a chemical that is used in dyeing when the dye will not fix directly onto the fabric. The mordant (mainly weak basic HYDROXIDES of aluminium, chromium and iron) impregnates the fabric, and the dye then reacts with the mordant, thus fixing it to the fabric.

morning star a planet (usually VENUS or possibly MERCURY) seen to the east in the sky around sunrise.

morphine a crystalline ALKALOID occurring in opium. It is used widely as a pain reliever but misuse can be dangerous.

Mössbauer effect the emission of resonant gamma radiation by excited nuclei that characterizes the chemical state of the nucleus. An effect used in the study of iron, tin and antimony compounds. It was discovered by the German physicist Rudolph Ludwig Mössbauer (1929–).

mother liquor the solution that remains after a substance has crystallized out of that solution.

motility the ability to move independently.

mucous membrane tissue that is found as a layer lining cavities in the body, which connect with the exterior, e.g. gut, respiratory tracts, etc. It is made up of an epithelium (a protective tissue, of closely packed cells, which has additional functions) that contains mucus-secreting cells and often CILIA.

mudflow the fluid movement of fine-grained sediment, which occurs typically in areas underlain by certain clays that can liquefy and flow (*see also* LAHAR). Mudflows are also associated with small tributary valleys in seasonal tropical climates, producing tongues and banks of unsorted debris.

multiple star a star system that contains three or more stars revolving around a common centre of gravity and held in position by their gravitational forces.

mutation a change, whether natural or artifical, spontaneous or induced, in the constitution of the DNA in CHROMOSOMES. Mutation is one way in which genetic variation occurs (*see* NATURAL SELECTION) since any change in the GAMETES may produce an inherited change in the characteristics of later generations of the organism. Mutations can be initiated by ionizing radiations and certain chemicals.

myelin sheath a fatty substance (**myelin**) that surrounds axons in the central nervous system of vertebrates and functions as an insulating layer.

mylonite a rock type produced in zones of tectonic dislocation, i.e. in fault and SHEAR ZONES. Rocks that are formed under "brittle" conditions through the breaking and crushing of material (cataclasis) may exhibit a FOLIATION, albeit crude, a grain size reduced by tectonic effect, and a hard, flinty texture, and are termed mylonites. If recrystallization has proceeded to the point where growth of new grains is dominant over grain size reduction, the rock is termed a blastomylonite. A mylonitic rock with random FABRIC is a cataclasite.

myoglobin an oxygen-carrying globular protein comprising a single POLYPEPTIDE chain and a haem group. The haem is an iron-containing molecule consisting essentially of a PORPHYRIN that holds an iron atom as a CHELATE (*see* CHELATION). The iron can reversibly bind oxygen in haemoglobin and myoglobin. The oxygen in myoglobin is released only at low oxygen concentrations, for example, during hard exercise when the muscles demand more oxygen than the blood can supply. Thus, myoglobin provides a secondary or emergency store of oxygen. Mammals such as whales have large amounts of myoglobin in their muscles to facilitate diving.

myosin a large protein found originally with ACTIN in muscles but occurring in most EUCARYOTIC cells. There are at least two varieties, one that is involved in cell locomotion and the other in muscle contraction. The myosin in muscles provides, with actin, the molecular basis of muscle contraction.

N

nacreous clouds a cloud formation at great height before sunrise or after sunset when its colouring is similar to mother of pearl.

nadir the lowest point—the pole that is vertically below the observer with the latter in the centre of the horizon circle. Opposite to the ZENITH in the CELESTIAL SPHERE.

nanometre a unit of measurement that is used for extremely small objects. One nanometre equals one thousand millionth of a metre.

naphtha a mixture of HYDROCARBONS obtained from several sources including COAL TAR and PETROLEUM. Naphtha from coal tar contains mainly AROMATIC hydrocarbons, whereas petroleum naphtha hydrocarbons are predominantly ALIPHATIC.

naphthalene consists of two condensed BENZENE rings, to give the formula $C_{10}H_8$. It is a white crystalline solid and is obtained from COAL TAR (in the fraction, or part, boiling between 180° and 200°C/356° and 392°F) and from PETROLEUM fractions, by demethylation, i.e. the removal of the METHYL group from methylnaphthalene. Naphthalene is more reactive than benzene, and it is used mostly in preparing the intermediate phthalic anhydride for the production of plasticizers and resins. Some derivatives of naphthalene are important in the manufacture of dyestuffs.

nappe a large-scale geological structure (tens of kilometres) occurring as a sheet of rock that has been displaced along a fault plane at its base. Many nappes have a THRUST at the base and are formed because of compression, but an alternative mechanism is the sliding due to gravity along a low-angle fault, as seen in the Swiss Alps. In many cases, the origin of nappes may be due to a combination of both mechanisms, which are difficult to extricate due to subsequent changes.

natural background (*see also* BACKGROUND) the radiation due to natural RADIOACTIVITY and COSMIC RAYS, which must be accounted for in the detection and measurement of radiation.

natural gas the description usually applied to gas associated with PETROLEUM, which originated from organic matter. The gas is a mixture of HYDROCARBONS, mainly methane and ethane with propane. Also present in small amounts may be butane, nitrogen, carbon dioxide and sulphur compounds (*see also* FOSSIL FUELS).

natural glass if MAGMA is cooled quickly, there is insufficient time for crystals to grow from the melt, and a glass is produced. The magma may be acidic or basic, in which case OBSIDIAN or tachylite will be produced. Glasses produced by meteoritic impact are found on the Moon. Glasses are METASTABLE solids.

natural logarithm *see* **logarithm**.

natural number any of the set of positive integers also known as the counting numbers: 1, 2, 3, 4 ....

natural selection the process by which evolutionary changes occur in organisms over a long period of time. Darwin explained natural selection (*see* DARWINISM) by arguing that organisms that are well adapted to their environment will survive to produce many offspring, whereas organisms that are not so well adapted to their enviromnent will not. As the better adapted organisms are successfully transmitting their genes from one generation to the next, these are the organisms that are "selected" to survive.

nautical mile the standard international unit of distance used in navigation. One nautical mile is defined as 1852 metres (6076 feet).

neap tides a TIDE of a range up to 30per cent less than the mean tidal range, which occurs near the Moon's first and third quarters (every 14 days) when the Moon, Earth and Sun are at right angles and the Sun's tidal influence works against the Moon's.

nebula (*plural* **nebulae**) clouds of interstellar medium, or a collection (GALAXY) of stars.

neck *see* **plug**.

nephoscope an apparatus for observing and measuring the speed of clouds and CELESTIAL BODIES, and their angular velocity.

Neptune (*see* APPENDIX 6 for physical data) the eighth planet of the SOLAR SYSTEM, with its orbit between URANUS and PLUTO. The surface temperature is approximately -200°C (-328°F), and the atmosphere consists mainly of METHANE, HYDROGEN and HELIUM. It has two satellites, Triton and Nereid.

neritic zone the shallow water marine zone near the shore that extends from low tide to a depth of approximately 200 metres (219 yards). Most BENTHIC organisms live in this zone because sunlight can penetrate to such depths. Sediments deposited here comprise sands and clays with features such as RIPPLE MARKS.

Nernst, Walther Hermann (1864-1941) a German physical chemist who was awarded the 1920 Nobel prize for chemistry for his proposal of the heat theorem, which was formulated as the third law of THERMODYNAMICS. In 1889, he also developed the Nernst equation, which is used to determine the electromotive force of a cell that contains non-standard constituents.

neuron another name for the nerve cell that is necessary for the transmission of information in the form of impulses along its body. The impulses are carried along long, thin structures of the neuron, known as the axon, and are received by shorter, more numerous structures called dendrites. It has been discovered that the transmission of impulses is faster in axons that are surrounded by MYELIN SHEATHS than those that are unmyelinated.

neutral a term indicating that a solution or substance is neither acidic or alkaline. The most well-known example of a neutral substance is pure water, which should have a pH of seven.

neutralization a reaction that either increases the pH of an acidic solution to neutral seven or decreases the pH of an alkaline solution to seven. Acid-base neutralizations occur in the presence of an INDICATOR, which undergoes a colour change when the reaction is complete.

neutrino *see* **elementary particle**.

neutron an uncharged particle that is found in the nucleus of an ATOM. The MASS of the neutron is 1.675

x 10^{-24}g, which is slightly larger than the mass of the
PROTON.

neutron diffraction the scattering of NEUTRONS by
the atoms in a crystal. Depending upon the nature
and structure of the crystal, information can be
gleaned about its atomic and magnetic structure.

neutron star a small body with a seemingly impos-
sibly high density. A star that has exhausted its fuel
supply collapses under gravitational forces so intense that its ELECTRONS and PROTONS are crushed
together and form NEUTRONS. This produces a star 10
million times more dense than a WHITE DWARF—
equivalent to a cupful of matter weighing many
million tons on Earth. Although no neutron stars
have definitely been identified, it is thought that
PULSARS may belong to this group.

névé *or* **firn** a compacted snow, which forms as
intermediate between snow and glacial ice, and
which has survived the melting of a summer.

newton the standard international unit of force.
One newton (N) is the force that gives one kg an
acceleration of one ms^{-2}.

Newtonian telescope a type of telescope in which
the image is reflected from a main mirror into an
eyepiece on the side of the tube.

Newton's laws of motion the fundamental laws of
mechanics, which describe the effects of force on
objects. Developed by Isaac Newton (1643–1727),
the famous English scientist, the three laws of
motion state:
First law—any object will remain in a state of rest
or constant linear motion provided no unbalanced
force acts upon it.
Second law—the rate of change of momentum is

proportional to the applied force and occurs in the linear direction in which the force acts.

Third law—every action has a reaction, which has a force equal in magnitude but opposite in its direction.

Until very recently, the above laws could not be directly demonstrated as they are idealized relationships that take place in the less idealized systems here on earth, where the effects of, say, FRICTION have to be considered.

niche all the environmental factors that affect an organism and its community. Such factors include spatial, dietary and physical conditions necessary for the survival and reproduction of a SPECIES. Only one species can occupy a specific niche within a community, as the coexistence of any species is subject to distinctions in their ecological niches.

Nicol prism a prism comprising two pieces of clear CALCITE cemented together by Canada balsam (a natural resin). Light entering the prism is doubly refracted and, on reaching the balsam cement, one ray is reflected away and the other carries on through the prisms, emerging as polarized light. Such prisms were incorporated in early versions of microscopes used to study thin sections of rocks, but discs of polaroid are now employed.

nimbostratus a grey layer of cloud that obscures the Sun and produces continuously falling rain (or snow).

nitration the addition of the nitro group (NO_2) to organic compounds. It is usually undertaken by reaction with nitric and sulphuric acids. It is used in the manufacture of explosives.

nitric acid a colourless, corrosive acid liquid, HNO_3. It is a powerful oxidizing agent, attacking most metals and producing nitrogen dioxide (NO_2). It is prepared on a large scale by passing a mixture of AMMONIA (NH_3) and air over heated platinum, which acts as a CATALYST. It is used widely in the chemical industry.

nitrides compounds of nitrogen with other elements. Nitrides are prepared by the action of nitrogen or gaseous ammonia. Electropositive elements (Group 1 and 2 metals) form ionic compounds, while elements from Groups 3 to 5 (*see* APPENDIX 1) form lattices with nitrogen in the holes. This results in compounds (refractory hard metals) that are hard, inert and have a high melting point.

nitrification the process by which soil bacteria convert ammonia (NH_3) present in decaying matter into nitrite (NO_2^-) and nitrate (NO_3^-) ions. The produced nitrate ions are taken up by plants and are used in their protein synthesis. Nitrification requires free oxygen, as the bacteria *Nitrosomonas* and *Nitrobacter* oxidize ammonia and nitrites respectively.

nitrogen a colourless, odourless gas, which exists as N_2 and forms almost 80 per cent of the atmosphere by volume. It is obtained commercially by fractionation of liquid air and is itself used extensively in producing AMMONIA, as a refrigerant, and as an inert atmosphere. It is a vital element for living organisms occurring in PROTEINS (and NUCLEIC ACIDS), and the NITROGEN CYCLE is the circulation of the element through plant and animal matter into and from the atmosphere.

nitrogenase the enzyme that catalyses the production of nitrogenous compounds from free NITROGEN in the process of NITROGEN FIXATION.

nitrogen cycle the regular circulation of NITROGEN due to the activity of organisms. Nitrogen is found in all living organisms and forms about 80 per cent of the atmosphere (this proportion is maintained by the nitrogen cycle). The start of the nitrogen cycle can be regarded as the uptake of free nitrogen in the atmosphere by bacteria (NITROGEN FIXATION) and the uptake of nitrate (NO_3^-) ions by plants. The nitrogen is incorporated into plant tissue, which in turn is eaten by animals. The nitrogen is returned to the soil by the decomposition of dead plants and animals. NITRIFICATION converts the decomposing matter into nitrate ions suitable for uptake by plants.

nitrogen fixation the process by which free nitrogen (N_2) is extracted from the atmosphere by certain bacteria. Some free-living bacteria can use the nitrogen to form their AMINO ACIDS, while other nitrogen-fixing bacteria live in the root nodules of leguminous plants (peas and beans) and provide the plants with nitrogenous products. This enables the plant to survive in nitrogen-poor conditions while the bacteria has access to a carbohydrate supply in the plant (a symbiotic relationship). The nitrogen-fixing bacteria are able to convert free nitrogen into nitrogenous products because of the presence of the enzyme nitrogenase within their cells.

noble gases the elements comprising group 8 of the PERIODIC TABLE. Noble gases are usually referred to as INERT GASES because of their relative unreactivity.

noble metals metals, such as platinum, silver and gold, that are highly resistant to attack by acids and corrosive compounds. They tend not to react chemically with non-metals.

noctilucent clouds thin, very high clouds of ice or

dust, which are sometimes brilliantly coloured due to their reflecting light from the Sun when it is below the horizon.

node the site on a plant stem where the bud and leaves arise. In astronomy, when the orbit of an object intersects a plane such as the equator, the two points of intersection are called nodes.

nodule a small swelling or structure on a plant, especially the root, which is due to nitrogen-fixing bacteria.

nomogram a chart comprising scaled lines to enable the value of a variable to be found, without calculation (derived from the plotting of known values from one or two other variables).

nonagon a nine-sided polygon that, if regular, has equal sides all with interior angles of 140°.

nondisjunction the failure of chromosomal pairs or sister CHROMATIDS to separate during MITOSIS or MEIOSIS respectively. Nondisjunction produces daughter cells containing an unequal number of CHROMOSOMES, either too many or too few.

non-ionic detergents detergents that do not ionize in water to produce large anions or cations (as soap does). Typical compounds are CONDENSATION products of glycols (dihydric alcohols with two hydroxyl groups in the molecule).

non-stoichiometric compounds compounds where the atoms present do not form simple ratios. Within a single crystal, departures from the ideal state may occur because of defects. The Frenkel defect is when an atom or ion is displaced from its regular position, and in the SCHOTTKY DEFECT, atoms or ions are actually missing. Hydrides of transition metals (*see* TRANSITION ELEMENTS) are good examples, and titanium hydride can form as a whole range of compositions depending upon the temperature and pressure of formation. It varies from $TiH_{0.1}$ to $TiH_{1.8}$.

normal a line perpendicular to the tangent of a curve or contact point of a line or plane.

normal fault a FAULT where the direction of dip is towards the downthrown side.

normal salt any SALT formed by an ACID, which loses more than one hydrogen ion (H^+) per molecule during a NEUTRALIZATION reaction. It should be noted, however, that the production of a normal salt is dependent on the quantities of the acid and BASE used in the reaction.

Northern Lights *see* **aurora**.

nova (*plural* **novae**) in the strict sense, a new star, but also a star that quite suddenly becomes brighter by a factor of five to ten thousand, due to an ejection of material in a gas cloud. It seems that novae are parts of BINARY STARS, and a WHITE DWARF is the source of the ejection or explosion.

nuclear chemistry the study of reactions involving the changes from one type of atom to another due to a nuclear reaction, achieved either by decay or collision with other particles.

nuclear fission a process that splits a heavy nucleus into two lighter nuclei. Nuclear fission produces more stable nuclei and emits huge amounts of energy. It is the process involved in the atom bomb, when uranium is bombarded with NEUTRONS, resulting in the splitting of uranium, which releases more neutrons, and a CHAIN REACTION ensues. Nuclear fission can be controlled by using synthetic rods to absorb excess neutrons. This is the basis of NUCLEAR POWER, which generates the energy used to propel nuclear submarines and to produce electricity from nuclear power plants.

nuclear fusion a process that fuses two lighter nuclei into a heavier, more stable nucleus. Nuclear fusion releases tremendous amounts of energy and is believed to be the energy source of the Sun as hydrogen nuclei are converted into helium nuclei. Very high temperatures are required for nuclear fusion, and the hydrogen bomb, which is based on fusion, needs the large energy source of NUCLEAR FISSION to initiate the explosion. Controlled nuclear fusion would be a greater energy source than controlled nuclear fission and would not generate radioactive byproducts, but, unfortunately, the byproduct of nuclear fusion—a high-temperature, dense gas—cannot, at present, be confined.

nuclear magnetic resonance spectroscopy an analytical technique based upon the absorption of electromagnetic radiation in the radio frequency region for those nuclei that spin about their own axes. The result is a change in the orientation of the spinning nuclei in a magnetic field. Of all the susceptible nuclei (^{1}H, ^{19}F, ^{31}P), ^{1}H is the most sensitive. The technique is used for the identification and analysis of organic materials, and magnetic resonance has been developed to form a medical imaging tool.

nuclear membrane a double membrane that surrounds the nucleus of EUCARYOTIC cells. There is a space between the outer and inner nuclear membranes, but both membranes seem to fuse together at regions containing nuclear pores. Nuclear pores contain proteins that probably control the exchange of material between the cell nucleus and its CYTOPLASM.

nuclear power the production of energy from the controlled NUCLEAR FISSION that involves uranium and plutonium as fuels. Nuclear power is used to generate electricity by removing the huge amount of energy released during fuel fission away from the core reactor to the outside, where it is converted to steam and generates electricity by driving turbines.

nucleic acid a linear MOLECULE that acts as the genetic information store of all cells. Nucleic acids occur in two forms, deoxyribonucleic acid (DNA) and ribonucleic acid (RNA), but both forms are composed of four different NUCLEOTIDES, which react to form the long chain-like molecule. DNA is found inside the nucleus of all EUCARYOTES as it is the major part of CHROMOSOMES, but RNA is found outside the nucleus and is essential for TRANSCRIPTION and TRANSLATION during protein synthesis.

nucleolus (*plural* **nucleoli**) a membrane-bound object found within the NUCLEUS. The nucleolus contains gene sequences that code for ribosomal RNA, ribosomal RNA itself, and proteins necessary for rRNA synthesis.

nucleophile a reactive molecule that will readily "donate" its unshared pair of electrons. Nucleophiles will attack the low electron density regions of other molecules.

nucleosynthesis a NUCLEAR FISSION reaction occurring in stars and supernovae explosions that produces elements other than hydrogen and helium.

nucleotide a MOLECULE that acts as the basic building block of the NUCLEIC ACIDS, DNA and RNA. The structure of a nucleotide can be divided into three parts—a five-carbon sugar molecule; a phosphate

271

group; and an organic base. The organic base can be either a PURINE, e.g. ADENINE, or a PYRIMIDINE, e.g CYTOSINE.

nucleus (*plural* **nuclei**) in biology, the ORGANELLE that contains the chromosomes of EUCARYOTIC cells. Molecules enter and leave the nucleus via the pores in the NUCLEAR MEMBRANE. Such molecules include AMINO ACIDS and MESSENGER RNA. In chemistry and physics, the term nucleus refers to the small, positively charged core of an ATOM that contains the PROTONS and NEUTRONS. The electrons of an atom orbit the nucleus.

In astronomy, the term has several uses including the central core of (of ice and dust) a COMET, and the central part of the galaxy.

nuée ardente a term, losing favour in current geological terminology, for a hot (or even incandescent) ash flow from a volcanic eruption. (It is now known to be a pyroclastic flow with ash cloud above). Such events are very dangerous because the flows can attain speeds of over 100 km/hr (62 mph). If the ash flow settles while hot, it forms a flattened, welded ash (*see also* IGNIMBRITE).

null hypothesis a hypothesis that examines the existence of a specific relationship by enabling it to be statistically tested. A null hypothesis assumes that the expected results from an experiment have just arisen from chance with no significant change occurring. These expected results are statistically compared with the observed results derived from the experiment. Any significant difference between the expected and the observed results will be taken as evidence to support the experimental hypothesis and to rule out the null hypothesis.

numerator the number or quantity to be divided by the denominator of a fraction. For example, the fraction $3/4$ has a numerator of 3.

numerical forecasting weather forecasting that relies upon numerous observations and measurements, both "on the ground" and in the atmosphere, and the subsequent calculation of what should follow (based also upon natural laws).

nunatak in an area undergoing glaciation, nunataks are rock mountain peaks that protrude above the ice.

O

obduction the stripping off of tectonic rock sequences
from lithosphere undergoing SUBDUCTION and their
pushing and thrusting onto the colliding plate. This
is due to the rocks, or terranes, being too buoyant to
be subducted.

oblique angle any angle that does not equal 90°
(right angle) or any multiple of 90°.

obsidian a glassy, silica-rich volcanic rock, formed
by the rapid cooling of a granitic melt. It is generally
black with a vitreous (glassy) lustre and conchoidal
FRACTURE.

occlusion the meeting of warm and cold air in a
DEPRESSION, where the warm air is lifted above the
colder air.

oceanic crust the upper part of the oceanic
lithosphere down to the MOHOROVCIC DISCONTINUITY.
It is formed of several layers, commencing with a
top layer of sediment that may be thin or absent (as
over oceanic ridges). However, near to continental
shelves, (*see* CONTINENTAL SHELF), sediment may accumulate in thicknesses up to 2 or 3 kilometres
(6562–9843 ft). Beneath this is a layer of LAVAS
(basaltic) and dykes, which together are about 2 km
thick. A gabbroic layer follows, approximately 5 km
(3 miles) thick, and the base of this layer approaches

the MANTLE in composition. These layers seem to remain remarkably constant between ocean basins (*see also* MID-OCEANIC RIDGE).

oceanic ridge *see* **mid-oceanic ridge**.

oceanography the study of the oceans, including tides, currents, the water and the sea floors.

obtuse angle any angle that lies between but does not equal 90° and 180°.

octahedron a geometrical solid that consists of eight planes, each bound by their edges. If the eight faces are all equilateral triangles, then the octahedron is said to be regular.

oestrogens a group of female sex hormones, including some sterols, e.g. oestradiol, which is one component of oral contraceptives.

offlap a sequence of inclined rock strata, with no breaks, deposited during the receding of marine waters such that successive layers are overlain laterally by younger rocks, marking the direction of the sea's retreat.

ohm (Ω) the unit of electrical resistance. Between two points of an electrical CONDUCTOR, one VOLT (V) is needed to force a current (I) of one AMPERE through a RESISTANCE (R) of one ohm, i.e. $V = IR$. This is known as **Ohm's law** (after the German physicist who formulated it, Georg Simon Ohm [1787–1854]), which can be rewritten: $R = V/I$.

oil a greasy liquid sustance obtained from animal or vegetable matter (*see* FATS) or from mineral matter (*see* PETROLEUM).There are essentially three groups of oils. The fixed or fatty oils from animal, vegetable and marine sources occur as glycerides and ESTERS. Mineral oils are HYDROCARBONS and are produced from petroleum, shale and coal, and essential oils are derived from certain plants and tend to be volatile hydrocarbons.

oil shale a dark, fine-grained shale (*see* SEDIMENTARY ROCK) containing organic substances that produce liquid HYDROCARBONS on heating, but do not contain free PETROLEUM.

olisthostrome a sedimentary deposit formed from a mixed and jumbled mass of large fragments of material older than the host sedimentary sequence. These sequences are formed by gravity sliding, sometimes into oceanic trenches, and are thus a feature of many SUBDUCTION zones.

olivine a rock-forming mineral with the formula $(Mg,Fe)_2SiO_4$. The olivine family exists as a SOLID SOLUTION between the end members forsterite (Mg) and fayalite (Fe). The mineral occurs in silica-poor IGNEOUS rocks (e.g. basalts and gabbro) and peridotite, which is a coarse-grained ULTRABASIC igneous rock with olivine, PYROXENE and GARNET. It is also found in STONY METEORITES and lunar basalts.

omnivore any organism that eats both plant and animal tissue.

oncogene any gene directly involved in cancer. Oncogenes may be part of a specific VIRUS that has managed to penetrate and replicate within the host's cell, or they may be part of the individual's GENOME, which has been transformed by radiation or a chemical.

ontogeny the complete development of an individual to maturity.

oocyte a cell that undergoes MEIOSIS to form the female reproductive cell (egg or ovum) of an organism. In humans, a newborn female already has primary oocytes, which will undergo further development when puberty is reached but which will only complete secondary meiosis to form the secondary HAPLOID oocyte if fertilization occurs.

oolite a LIMESTONE made up largely of ooliths (or ooids). Ooliths are pellet-like growths of CALCIUM CARBONATE. They have a fragment of, for example, shell as the core, and this is surrounded by concentric layers of calcium carbonate. Their formation is favoured by shallow marine conditions with raised temperature and salinity.

ooze a deep sea mud made up of clays and the calcareous or siliceous remains of certain organisms, e.g. diatoms.

opalescence a term used in mineralogy to describe the pearly lustre that resembles opal. It is caused by the interference of light being reflected and refracted from very thin layers at the surface of the mineral. Opal contains water, which creates this effect, but water may be lost if the mineral is left exposed to the air and the play of colours may disappear.

open chain a compound with an open chain not a ring structure, as in aliphatic compounds, e.g. ALKANES, ALKENES and ALKYNES and their derivatives.

open clusters clusters of loosely gathered stars that have a similar motion through space. There are likely to be a few hundred stars, with gas and dust clouds, e.g. the Ursa Major cluster.

optical activity *or* **optical rotation** is when a solution of a substance rotates the plane of transmitted polarized light (*see* POLARIZATION). It occurs with optical ISOMERS, with one form rotating the light in one direction and the other rotating the light by the same amount in the opposite direction

(DEXTROROTATORY and LAEVOROTATORY compounds). A racemic mixture is optically inactive because it contains equal amounts of both forms.

optical fibre a small, thin strand of pure glass that uses internal reflection to transmit light signals. Optical fibres are more efficient than conventional cables as they are much smaller, thus requiring less space, and have a higher data-carrying capacity.

optical isomerism the existence of two chemical compounds that are ISOMERS, which form non-superimposable mirror images.

optical rotation *see* **optical activity**.

orbit the path of a heavenly body (and satellite or spacecraft) moving around another due to gravitational attraction.

orbitals orbitals are a means of expressing, rationalizing and correlating atomic structure, bonding and similar phenomena. The BOHR THEORY postulated the positioning of ELECTRONS in definite orbits about a central NUCLEUS. However, it was soon discovered that this was too simple and that electrons behave in some ways as waves, which makes their spatial position more imprecise. Hence the old "particle in an orbit" picture was replaced by an electron "smeared out" into a charge cloud or orbital, which represents the probability distribution of the electron. An atomic orbital is thus one associated with an atomic nucleus and has a shape determined by QUANTUM NUMBERS. Various types of orbital, designated s, p, d, etc, are distinguished. An s orbital is spherical, and a p orbital is dumbbell-shaped. When two atoms form a COVALENT BOND, a molecular orbital with two electrons is formed, associated with both nuclei (*see* σ and π BOND). The overlapping of atomic orbitals in a carbon-carbon single bond (e.g. ETHANE) creates a molecular orbital centred on the line joining the two nuclei. In a carbon-carbon double bond (e.g. ETHENE), the second of the two bonds is created by two overlapping p orbitals, forming the π bond. The two overlapping dumbbells create two sausage-like spaces of electron "cloud" on each side of the line joining the nuclei. BENZENE has torus-shaped (doughnut-shaped) molecular orbitals on each side of the ring due to overlap and merging of p atomic orbitals.

order of magnitude the approximate size of an object or quantity usually expressed in powers of 10.

ordinal number in general, 1st, 2nd, etc. rather than 1, 2, etc. (cardinal numbers).

ordinal scale a statistical scale that arranges the data in order of rank in the absence of a numerical scale with regular intervals. Ordinal scales are ideal for data that contain relationships such as bad, good, better, best, as the data can be put in rank order but no regular interval can be measured between the ranked judgements.

ordinate the vertical or y-axis in a geometrical diagram for CARTESIAN CO-ORDINATES. For example, a point with co-ordinates (2,-6) has an ordinate of -6.

ore any naturally occurring substance that contains commercially useful metals or other compounds. The extraction of the desired metal will only proceed if the process is both economically and chemically feasible. Some relatively unreactive metals, such as copper and gold, exist as native ores with no need of extraction, but most metals are obtained by extracting them from their oxygen-containing (oxide) ores.

organelle any functional entity that is bound by a membrane to separate it from the other cell constituents. Organelles are found in the cells of all EUCARYOTES and include the CHLOROPLASTS and vacuoles of plant cells in addition to the NUCLEUS, MITOCHONDRIA, GOLGI APPARATUS, ENDOPLASMIC RETICULUM, and other small vesicles of both animal and plant cells.

organic chemistry the branch of chemistry concerned with the study of carbon compounds. Organic chemistry studies the typical bond arrangements and properties of carbon compounds containing hydrogen and, less frequently, oxygen and nitrogen. As most organic compounds are derived from living organisms, two major areas for study are the biologically important organic compounds and the commercially important organic compounds, e.g. ALKANES derived from oil.

origin the point of intersection of the horizontal (x-axis) and the vertical (y-axis) axes in a two-dimensional diagram for CARTESIAN CO-ORDINATES. Thus the origin has co-ordinates of (0,0).

orogeny a period of mountain building, many of which have occurred in the past, each lasting for millions of years (*see* MOBILE BELT).

orographic ascent the movement of air upwards when blowing over a mountain. Orographic cloud and rain are produced by the forcing of moist air upwards, which is then cooled, resulting in condensation.

oscillation the regular production of a fluctuating position or state. Oscillation can mean the cycle, vibration or rotation of the object in question. In the case of a simple pendulum, oscillation refers to its regular swinging motion and, when used in connection with electrical circuits, oscillation refers to the production of an ALTERNATING CURRENT.

oscillator a CIRCUIT or device that produces an ALTERNATING CURRENT or voltage of a specific FREQUENCY as its output signal. Two necessary components of an oscillator are the capacitor, which produces the electric field, and the inductor (coil), which produces the magnetic field (*see* ELECTROMAGNETIC INDUCTION). The alternating current of an oscillator is fundamentally caused by energy transfers between the electrical energy stored in the capacitor and the magnetic energy stored in the inductor. The capacitor also determines the frequency, as its value can be changed to get the desired frequency.

osmosis the process in which solvent molecules (usually water) move through a semi-permeable membrane to the more concentrated solution. Many mechanisms have evolved to prevent the death of animal cells either by too much water entering a cell by osmosis, causing it to rupture, or by too much leaving by osmosis, causing it to shrink (plasmolysis). Such mechanisms include the presence of a pump within the membrane of animal cells, which actively regulates the concentration of vital cellular IONS and the excretion of salt through the gills of marine bony fish to remove the salt gained by diffusion and drinking.

outcrop exposure at the Earth's surface of part of a rock formation.

outlier a feature created when an outcrop of younger rocks is completely surrounded by older rocks. It is usually produced by erosion of a folded or faulted rock sequence.

outwash fan a fan-shaped deposit of sands and gravel laid down by glacial melt-water at the margin of BOULDER CLAY derived from the glacier.

overburden within a sequence of sediment, the strata that lie over and therefore compress those beneath. Also, superficial material that overlies solid rock.

overlap a deposition of sediments caused by an advance of the sea to cover more and more land and deposit more sediments. As the sea advances (transgression), progressively younger sediments are laid down over and beyond older ones, creating an UNCONFORMITY.

overstep is where a series of rocks rests upon progressively older rocks beneath an unconformity, suggesting the older rocks are tilted.

oviparous a term describing animal reproduction where the development of the embryo and subsequent hatching occurs outside the female's body. Oviparous reproduction is found in birds, most fish, and reptiles.

ovoviviparous the term to describe the development of offspring within the body of the female in the absence of a placenta. Ovoviviparous reproduction is found in certain species of fish, reptiles and insects, where the young are retained within the mother's body solely for protection as they receive their nutrients from the egg and not from the mother.

oxbow lake the development of river MEANDERS into large loops. Eventually the "neck" between a looped meander is cut and the river straightens its course, leaving a cut-off loop or oxbow lake.

oxidation any chemical reaction that is characterized by the gain of OXYGEN or the loss of ELECTRONS from the reactant. Oxidation can occur in the absence of oxygen, as a molecule is also said to be oxidized if it loses a hydrogen atom.

oxide a compound formed by the combination of OXYGEN with other elements, with the exception of the INERT GASES.

oxidizing agent any substance that will gain ELECTRONS during a chemical reaction. Oxidizing agents will readily cause the OXIDATION of other atoms, molecules or compounds, depending on the strength of the oxidizing agent and the reactivity of the other substance. The following are all examples of oxidizing agents arranged in order of increasing oxidizing strength—sodium ions (Na^+), sulphate ions (SO_4^{2-}), and oxygen molecules (O_2).

oxygen a colourless and odourless gas, which is essential for the respiration of most life forms. It is the most abundant of all the elements, forming 20 per cent by volume of the atmosphere; about 90 per cent by weight of water; and 50 per cent by weight of ROCKS in the crust. It is manufactured by the FRACTIONAL DISTILLATION of liquid air and is used for welding, anaesthesia and rocket fuels.

ozone a denser form of oxygen that exists as three atoms per molecule (O_3). Ozone is a more reactive gas than the more common diatomic molecule (O_2), and can react with some hydrocarbons in the presence of sunlight to produce toxic substances that are irritants to the eyes, skin and lungs. Minute quantities of O_3 are found in sea water. It forms the earth's OZONE LAYER, 15 to 30 kilometres (9 to 18 miles) above the earth's surface.

ozone layer a region of the earth's atmosphere containing ozone that acts as a barrier against the

ULTRAVIOLET RADIATION from the sun. Scientists and environmentalists have announced that large holes are appearing in the ozone layer as a result of the widespread use of the ozone-depleting chemicals called chlorofluorocarbons (CFCs). CFCs are used in many industrial processes and, because of their unreactive qualities, as aerosol propellants. In the earth's atmosphere, however, they will readily react with, and destroy, ozone in the presence of sunlight. The depletion of the ozone layer is cause for concern as increased exposure to ultraviolet radiation will increase the incidence of skin cancers and eye cataracts. To prevent further damage to the ozone layer, industrial nations are being called upon to greatly reduce the use of CFCs.

P

palaeocurrent the preservation of a current indicator in sedimentary rocks, whether the original feature was due to ice, water or winds. Study of the directional data derived from ripples, bars, channels, etc, can provide information on current flow and sediment movement across a sedimentary basin.

palaeogeography the study of the physical geography at periods in the past.

palaeontology the scientific study of FOSSILS.

palynology the study of (mainly fossil) pollen, spores and some other microfossils. It deals primarily with structure, classification and distribution, and is useful in petroleum exploration and palaeoclimatology. In the main, the spores and pollen are highly resistant, and in some sedimentary sequences are the only fossils that can be used for stratigraphic correlation.

Pangaea (*see also* CONTINENTAL DRIFT) the single continent, proposed by the German meteorologist and geophysicist Alfred Wegener (1880–1930) in 1915, which came into being late in the Permian and lasted almost to the end of the Triassic. The construction of this "supercontinent" was supported by the geometric fit of coastlines on either side of the

Atlantic, the similarity between faunal distributions, and other factors.

parabola a plane curve traced out by a point moving so that its distance from a fixed point (focus) is equal to its perpendicular distance from a fixed straight line (directrix).

parabolic rule *see* **Simpson's rule**.

parabolic velocity in astronomy, the velocity a body would need to make a parabola about the centre of attraction (also called ESCAPE VELOCITY).

paraffins (*see* ALKANE) the general/informal term for saturated ALIPHATIC HYDROCARBONS with the formula C_nH_{2n+2}. They are quite unreactive, hence the name paraffin (from the Latin *parum affinis*, "little allied").

parallax the apparent movement in the position of a celestial object due to a change in the position of the observer and therefore due in reality to the Earth moving through space.

parallel circuit a CIRCUIT in which each component has the same POTENTIAL DIFFERENCE but has a different amount of current flowing through it. The amount of current flowing through each component depends upon a phenomenon called RESISTANCE, and in parallel circuits the total resistance (R) of any circuit components is given by the following relationship:

$$1/R = 1/R1 + 1/R2 + 1/R3 \ldots \text{etc.}$$

parallelogram a four-sided POLYGON, which has opposite sides that are parallel and equal in length. Parallelograms can have four sides all of equal length (equilateral parallelogram, i.e. a rhombus), four equal angles (equiangular parallelogram, i.e. a rectangle) or have all four sides and angles equal, i.e. a square.

parameter an arbitrary constant or variable that determines the specific form of a mathematical equation, as a and b in $y = (x - a)^2 + b$. Changing the value of the parameter generates various cases of the phenomena represented.

parametric equation an equation where the co-ordinates of a point appear dependent on parameters. Any point (x, y) on a curve or on a surface may be expressed as functions of a third variable, t, such that x and y are functions of t; $x = f(t)$, $y = g(t)$.

parasite any organism that obtains its nutrients by living in or on the body of another organism (its host). Parasites can be either completely dependent on their host for survival (obligate) or are able to survive without their host (facultative). The extent of the damage on the health of the host by parasitic infestation can range from being virtually harmless to so severe that it causes the death of the host. Highly evolved parasites are so well adapted, however, that the host is able to survive and reproduce as normal, thus providing the parasite with a permanent supply of nutrients.

parenthesis (*plural* **parentheses**) the curved brackets ( ) used to group terms or as a sign of aggregation in a mathematical or logical expression.

parsec an astronomical unit of distance, which is used for measurements beyond the solar system and which corresponds to a PARALLAX of one second of arc. It equals 206,265 ASTRONOMICAL UNITS or 3.26 light years.

parthenogenesis the development of a new individual from an unfertilized egg. Parthenogenesis is most common among the lower invertebrates, such as insects and flatworms. For example, the process can be part of the honey bee life cycle if the HAPLOID eggs laid by the queen remain unfertilized by sperm. The larvae from these eggs will develop into the male bees (drones), whose only function is to produce sperm. If the eggs laid by the queen are fertilized then the larvae develop into sterile female worker bees or fertile queens, depending on the food supply. Parthenogenesis also occurs in some plants, such as the common dandelion, and it can be induced artificially in many species by stimulation of the egg cell.

partial fractions the simple FRACTIONS into which a larger fraction may be separated so that the sum of the simpler fractions equals the original larger fraction.

particle the concept in physics of a small body that has a finite mass but insignificant dimensions, so that it has no moment of inertia about its centre of mass.

particle shape *or* **grain shape** the shape of a particle is described by reference to the dimensions along three mutually perpendicular axes within a grain. There are four shape classes defined by dimension ratios: oblate (tabular or disc-shaped); prolate (like a rod); bladed; and equant (i.e. equal as in cube or sphere).

particle size *or* **grain size** the diameter of the grains in a SEDIMENTARY ROCK. The size is determined by physical means—sieving for the smaller grains, and direct measurement for the larger. It is customary to size small grains by referring to the diameter of a sphere with the same volume. Two commonly used classifications are the Udden-Wentworth scale and the British Standard classification (BSI):

Udden-Wentworth		*BSI*	
> 256 mm	boulder	> 200 mm	boulder
64–256 mm	cobble	60–200 mm	cobble
2–64 mm	pebble	2–60 mm	gravel
62.5–2000 µm	sand	600–2000 µm	coarse sand
4–62.5 µm	silt	200–600 µm	medium sand
< 4 µm	clay	60–200 µm	fine sand
		2–60 µm	silt
		< 2 µm	clay

pascal the unit of PRESSURE named after the French philosopher and physicist, Blaise Pascal (1623-1662). One pascal (Pa) is defined as the FORCE of one NEWTON acting on a square metre, i.e. $1 \text{ Pa} = 1 \text{ Nm}^{-2}$. One atmosphere pressure (760 mm Hg, the air pressure at sea level) is approximately 100 kilopascals (kPa).

Pascal's law of fluid pressures the PRESSURE of a fluid is the same at any point since any applied pressure will be transmitted equally to all points of the containing vessel. Pascal discovered this principle while mountaineering with his father. He realized that the column of mercury in the barometers he carried would vary in length—essentially the principle behind all hydraulic systems.

Pascal's triangle the diagrammatical array of integers starting with one such that each number is the sum of the two numbers in the row directly above it. The result is a triangle of potentially infinite size, the beginning of which is as follows:

```
            1
          1   1
        1   2   1
      1   3   3   1
    1   4   6   4   1
  1   5  10  10   5   1
```

Pascal's triangle is an extremely useful method for determining the COEFFICIENTS when using the BINOMIAL THEOREM for expanding equations of (a + b)n, where the nth line of the triangle corresponds to n.

passive immunity the ability of an individual to resist disease using ANTIBODIES that have been donated by another individual rather than by producing its own antibodies. Passive immunity is obtained by young mammals from their mother's milk during the first few weeks of life as the newly born are virtually incapable of antibody production. In humans, breast-fed infants will receive most of their maternal antibodies from their mother's milk, but one antibody, IgG, will be found in all infants, whether breast-fed or bottle-fed, as IgG can cross the placenta during foetal development.

passive margin continental margins that lie within plates, e.g. the Atlantic margin of America. These areas tend to have thick sequences of SEDIMENTARY ROCKS, often rifted, and can be suitable locations for oil and gas.

Pasteur, Louis (1822-1895) French chemist and bacteriologist who was the first to demonstrate that a colony could be grown in a culture medium that had been infected with a few cells of the microorganism. This experiment showed that living cells had an inheritance of their own and helped discredit the theory of SPONTANEOUS GENERATION of life. Although Pasteur had been aware of the role of microorganisms in FERMENTATION since 1858, he did not accept their role in causing disease until several years later. He demonstrated that attenuated forms of micro-organisms could be used in innoculation, providing immunization for the host, and in

1885 he produced the first rabies vaccine.

pasteurization a process developed by PASTEUR of partially sterilizing food by heating it to a certain temperature. Food is pasteurized before distribution as the process can destroy potentially harmful bacteria, e.g. heating milk for 30 minutes at 62°C destroys the bacteria responsible for tuberculosis, and increases the shelf life of food by delaying its FERMENTATION.

pathogen any organism that causes disease in another organism. Most pathogens that affect humans and other animals are bacteria or viruses, but in plants there is also a wide range of fungi that act as pathogens.

Pauling, Linus Carl (1901-) an American biochemist who determined the structure of crystals of simple molecules and pure proteins by using X-ray crystallography. With fellow colleagues, Pauling discovered one of the regular structures common to all proteins, the α-helix and, using ELECTROPHORESIS, isolated the abnormal HAEMOGLOBIN that causes an hereditary form of anaemia. He also devised the Pauling scale, which is a useful method of making qualitative comparisons between the ELECTRONEGA-TIVITIES of the elements.

peat an organic deposit formed from compacted dead and possibly altered, vegetation. It is formed from vegetation in swampy hollows and occurs when decomposition is slow because of the ANAEROBIC conditions in the waterlogged hollow. Sphagnum moss is among the principal source-plants for peat. As peat builds up each year, water is squeezed out of the lower layers, causing the peat to shrink and consolidate. Even so, cut peat has a high moisture content and is dried in air before burning. In Ireland and Sweden, peat is used in power stations.

pectins complex POLYSACCHARIDES occurring in the cell walls of certain plants, particularly fruits. They are soluble in water and acid solutions, and gel with sucrose, i.e. they set to a jelly, hence their use in jam making.

pedology the study of soils—their composition, occurrence and formation.

peduncle in botany, the main stalk of a plant bearing several flowers. In zoology, the stalk by which certain organisms, e.g. brachiopods, anchor themselves to the substrate.

pegmatite a very coarse-grained IGNEOUS rock of essentially granitic composition. Pegmatites are characterized by the growth of large crystals due to them being late-stage crystallization products of MAGMA, enriched in VOLATILES and trace elements. As such, pegmatites often concentrate rare elements, and, in addition to alkali FELDSPAR, QUARTZ, and MICA, may produce BERYLS, topaz and tourmaline. The scale can be impressive, with sheets of mica many centimetres across, beryl as thick as branches, feldspars the size of large boxes, and on occasion much larger! Most pegmatites occur as veins, DYKES or lenticular bodies.

pelagic said of organisms living in the sea between the surface and middle depths. Pelagic sediments (e.g. OOZE) are deep-water deposits comprising minute organisms and small quantities of fine-grained debris.

pelitic rock a METAMORPHIC ROCK formed by the metamorphism of shales and mudstones. The particular aluminium-silicate minerals formed will depend on the conditions of metamorphism but usually include MICA (*see* BARROVIAN METAMORPHISM).

pendulum a device comprising a weight swinging on the end of a wire of negligible mass suspended from a fixed point. The time of a complete swing is given as

$$T = 2\pi\sqrt{(l/g)}$$

where l is the length of the wire and g the acceleration due to gravity.

penecontemporaneous a term used to describe any process happening in a rock soon after its formation.

peneplain the final stage in an erosional cycle (*see* MATURITY) when a region is worn down to an undulating plain characterized by low relief, with small hills and wide, shallow river valleys. Residual hills may remain, and these are called monadnocks, after Mount Monadnock in New Hampshire, USA.

pentad a period of five days, which is used in meteorological records because it is a fraction of a normal year.

pentagon any plane shape that has five sides. A regular pentagon has sides of equal length and five interior angles each measuring 108°.

pentahedron any three-dimensional figure that has five plane faces.

pepsin a proteolytic (*see* PROTEOLYSIS) enzyme that is secreted by cells lining the stomach. It is active in acid conditions and catalyses the breakdown of proteins, giving PEPTIDEs and AMINO ACIDs.

peptide bond the chemical linkage formed when two AMINO ACIDs join together. As all amino acids have a common molecular structure, the reaction always involves the elimination of a water molecule as the amino group (NH_2) of one amino acid molecule joins to the carboxyl group (COOH) of another molecule.

periastron when a body is orbiting a star, the periastron is the point at which the body is nearest to the star.

perigee *see* **apogee**.

perihelion similar to PERIASTRON but referring to the Sun, *viz* the point when a body's orbit takes it closest to the Sun. This applies to planets, comets, space-craft, etc.

perimeter the total distance round the outside of a closed plane figure, such as the circumference of a circle.

period the time taken for a body to complete one full OSCILLATION, which can involve vibrational, rotational or harmonic motion. Period (T) has seconds (s) as its units, and it is the reciprocal of FREQUENCY, i.e. $T = 1/f$.

In chemistry, the horizontal rows in the PERIODIC TABLE i.e. those elements between an alkali metal and the next inert gas. The periods are thus hydrogen (H) and helium (He); two periods of few elements, Lithium (Li) to Neon (Ne), and Sodium (Na) to Argon (Ar); the two long periods containing the TRANSITION ELEMENTS, running from Potassium (K) to Krypton (Kr) and Rubidium (Rb) to Xenon (Xe); the period Caesium (Cs) to Radon (Rn) and the unfinished period beginning with Francium (Fr). A geological period is the second order of geological time e.g. the Carboniferous Period (*see* APPENDIX 5).

periodic function a mathematical function (e.g. sine or cosine) whose possible values all recur at regular intervals. The graph of the function $y = \sin x$, where x is the number of degrees, produces a curve that repeats itself every 360°, i.e. it has a period of 360°. In general, a function $f(x)$ of a real or complex variable is periodic, with period T if f(x + T) = f(x) for every value of x.

periodic table an ordered arrangement of the elements by their ATOMIC NUMBER. The elements are arranged by PERIODS (horizontally, *see* APPENDIX 1), which correspond to the filling of successive shells, and by groups (vertically), which reflect the number of VALENCY ELECTRONS, i.e. the number in the outer shell.

period of revolution the average observational value for one complete revolution of a planet around the Sun, or a satellite around a planet.

peristalsis the involuntary muscular contractions responsible for moving the contents of tubular organs in one direction. Peristalsis occurs in the alimentary canal of animals as the alternate waves of contraction and relaxation of smooth muscle move food and waste products along.

perlitic structure a texture found in glassy or devitrified (*see* DEVITRIFICATION) igneous glasses, seen as curved or sub-spherical cracks. It is formed due to the contraction of the MAGMA during rapid cooling.

permafrost ground that is permanently frozen save for surface melting in the summer. About one quarter of the Earth's land surface is affected, and although it may be very thick (several hundred metres) the larger depths are probably relics from the last ice age. It occurs north of the Arctic Circle in Canada, Alaska and Siberia. Permafrost can cause considerable engineering problems, particularly when the heat generated by towns, etc, creates some thawing, leading to subsidence and slumping of previously solid ground.

permeability the ability of a sediment, soil or rock

to allow the flow of fluids (here taken as gas, oil or water). Specifically, the permeability (or hydraulic conductivity) is the volume flow rate of water through a section of porous medium under the effect of an hydraulic gradient at a certain temperature.

In physics, it is the diffusion rate of a liquid or gas through a porous material, under the effects of a pressure gradient.

permutation an ordered arrangement of a set of objects into specified groups. The number and order of component objects is important, e.g. the arrangement of four letters, ABCD, taken two at a time yields 12 permutations: AB, AC, AD, BC, BD, CD, BA, CA, DA, CB, DB, DC. The formula for the number of permutations that can be made from n dissimilar objects taken r at a time is n!/(n − r)!, where ! stands for FACTORIAL.

perpendicular any line or plane that meets another line or plane at a right angle (90°). If the perpendicular is formed by a line meeting a plane or the tangent to a curve, then the line is referred to as the NORMAL of that plane or curve.

perturbation slight changes in the movement of planets from their orbits because of gravitational attraction, drag, etc.

petrifaction the process by which organic remains are replaced by, usually, minerals, but in which the original structure is retained (*see also* FOSSILIZATION).

petrography the description of rocks, the minerals present and the textures, through study of hand specimens and THIN SECTIONS (under the microscope).

petroleum *or* **crude oil** a mixture of naturally occurring HYDROCARBONS formed by the decay of organic matter under pressure and elevated temperatures. Oil thus formed migrates from its source to a permeable reservoir rock, which is capped or sealed by an impermeable cover. The composition of the petroleum varies with the source and is separated initially by FRACTIONAL DISTILLATION into its major components (gas, liquids, wax, and residues such as bitumen). The liquids include petrol, paraffin oil, and other hydrocarbon liquids. CRACKING is used to break down some substances to create smaller molecules that can be used more readily. In addition to the production of various fuels, petroleum is the basis of the vast petrochemicals industry. *See also* FOSSIL FUELS.

petrology the study of rocks, in particular their origin, occurrence, mineral and chemical composition, and any processes of alteration. The term is usually applied separately to each "family" of rock type, i.e. sedimentary, igneous or metamorphic petrology.

pH the measure of concentration of hydrogen IONS (H^+) in an aqueous solution. The pH is the negative LOGARITHM (base 10) of H^+ ion concentration, calculated using the following formula:

$$pH = \log_{10} (1/(H^+))$$

The scale of pH ranges from 1.0 (highly acidic), with decreasing acidity until pH 7.0 (NEUTRAL) and then increasing alkalinity to 14 (highly alkaline). As the pH measurement is logarithmic, one unit of pH change is equivalent to a tenfold change in the concentration of H^+ ions.

phagocytosis the process by which cells bind and ingest large particles from the surrounding environment. In phagocytosis, the target particle binds to the cell's surface and is then completely engulfed by a bud formed by the plasma membrane of the cell. This process is used by simple unicellular organisms to ingest food particles and by certain LEUCOCYTES to engulf and destroy bacteria and old, broken cells.

phase applied to the MOON, the term used for the change in shape of the bright surface due to the relative positions of the Sun, Moon and Earth.

A chemical phase is a part of a system that is chemically and physically uniform and that is separated from other such parts by boundary surfaces, i.e. a distinct interface. Thus, in a system with ice, water and water vapour, there are three phases. However, all gases show one phase since they all mix with each other.

phase rule *or* **Gibbs phase rule** a relationship devised by the American chemist Josiah Willard Gibbs (1839–1903) relating number of phases (P), to degrees of freedom (F) and the number of components (C). Thus in a chemical system

$$P + F = C + 2$$

phenol (C_6H_5OH—carbolic acid) as a solution in water, it is corrosive and poisonous. It is used as a disinfectant (with the typical "carbolic" smell) and in the manufacture of dyes and PLASTICS.

phenocrysts large crystals, usually well-formed, that occur in a GROUNDMASS of smaller crystals. An IGNEOUS ROCK showing two sizes of crystals in this way is said to be porphyritic.

phenotype the detectable characteristics of an organism that are determined by the interaction between its GENOTYPE and the environment in which the organism develops. Organisms with identical genotypes may have different phenotypes, due to development in environments that differ in, for example, the availability of important nutrients or specific stimuli. It is unlikely, however, that organisms that have identical detectable phenotypes will have different genotypes unless they are HETEROZYGOTES. The expression of the dominant gene masks the presence of a recessive gene, as only the expressed gene affects the organism's phenotype.

pheromone a molecule that functions as a chemical communication signal between individuals of the same species. Pheromones are used extensively throughout the animal kingdom and have a wide range of functions. They can act as sexual attractants (very common in insects) and can help establish territories, as demonstrated by the frequent urination by dogs. Although pheromones are much rarer in plants than animals, one of the most economically and environmentally important pheromones is produced by a plant called the "Scary Hairy Wild Potato" (*Solanum berthaultii*). The leaves of this plant produce a pheromone that is identical to the warning signal produced by aphids. Breeding this aphid-repellent character into cultivated crops will reduce the financial loss from crop damage and reduce pollution as insecticides are needed less.

phloem tissue in plants that has the major task of transporting food materials from the points of production, the leaves, to areas where they are needed, e.g. growing points. The phloem is made up of sieve tubes, which are hollow and lie parallel to the length of the plant. The tubes are formed from sieve cells, end to end, with end cell walls broken down to permit movement of the metabolites.

phonon the QUANTUM of heat energy in the lattice vibrations of a crystal.

phospholipids lipids containing phosphoric acid (H_3PO_4) groups and nitrogenous bases. Phospholipids are found in brain tissue and egg yolks.

phosphorescence LUMINESCENCE that continues after the initial cause of excitation. The substance usually emits light of a particular WAVELENGTH after absorbing ELECTROMAGNETIC radiation of a shorter wavelength.

phosphorus a non-metallic element (P) occurring in several allotropic (*see* ALLOTROPY) forms: red, white and black, the latter being a high-temperature and pressure variety. Phosphorus occurs naturally as compounds, mainly as calcium phosphate ($Ca_3(PO_4)_2$). It is manufactured by heating the phosphate with sand and carbon in an electric furnace. It is commonly found in minerals and living matter, and is vital to life, being the main constituent of animal bones. It is used industrially in the manufacture of fertilizers, matches and in organic synthesis.

photic zone the uppermost layer of a lake or sea where there is adequate light to allow PHOTOSYNTHESIS to proceed. The limit will vary on the quality of the water and the material held in suspension, but can be as much as 200 metres (656 ft).

photobiology the study of the effect of light on living organisms.

photochemistry the study of the effects of radiation (mainly the visible and ultraviolet parts of the SPECTRUM) on chemical reactions. Only light that is absorbed can have any effect, and the first stage in a photochemical reaction is the absorption of (a quantum of) light energy by an atom, which is then raised to an excited state. Infrared radiation is ineffective, but radiation from the far ultraviolet is strong enough to break chemical bonds. The light absorbed may catalyse a reaction (*see* CATALYST) or render possible a reaction that would otherwise not proceed. A reactant usually absorbs the light, but where the energy is passed on to a reactant by another species, the process is called photosensitization.

photochromics the term used for materials that are sensitive to light. In some cases materials darken in bright light, and the change is reversed when the light source is removed. Certain materials are used in optical memory devices. The phenomenon is called photochromism, or phototropism.

photolysis the decomposition or reaction of a substance due to the absorption of light or ultraviolet radiation. Flash photolysis is a technique for studying very fast reactions involving ATOMS or RADICALS in the gas phase. The reactants are subjected to an intense, but brief, flash of light, which causes dissociation. Subsequent flashes are used immediately afterwards to identify intermediates produced in the reaction (by studying the absorption spectra).

photometry the measurement of electromagnetic energy (usually light) received from a celestial object. Similarly in physics, the measurement of the intensity of light sources.

photosphere the Sun's visible surface, which is several hundred kilometres thick and which is estimated to have a temperature of 6000K. Sunspots manifest themselves on this surface.

phototaxis the movement or reaction of an organism in response to light (*see also* CHEMOTAXIS).

photon a QUANTUM of energy that is an intrinsic component of all ELECTROMAGNETIC WAVES. Photons are used to explain the quantum theory of light, where the properties of light are explained in terms of particles (photons), as opposed to the wave theory of light, where its properties are explained by the propagation of a wave and how it disturbs a medium. The energy of a photon is proportional to the FREQUENCY of the light beam.

photosynthesis the process by which plants make carbohydrates, using water, carbon dioxide (CO_2) and light energy, while releasing oxygen. Photosynthesis occurs in two stages, known as the CALVIN CYCLE and the LIGHT REACTIONS of photosynthesis. For photosynthesis to occur, an organism must contain light-trapping pigments, which capture light energy in the form of PHOTONS and use the photons to initiate a series of energy-transfer reactions. Some blue-green algae (cyanobacteria) and the CHLOROPLASTS of all plants contain the essential light-trapping pigment called CHLOROPHYLL that makes them capable of photosynthesis. Photosynthesis is an essential process for regulating the atmosphere as it increases the oxygen concentration while reducing the CO_2 concentration.

phototropism a growth movement exhibited by parts of plants in response to the stimulus of light. Plant shoots display positive phototropism as they grow towards the light source, but the roots tend to display negative phototropism as they grow away from the light source (*see* GEOTROPISM). Phototropism is caused by the unequal distribution of auxin (a plant growth hormone) as this substance has a higher concentration in the darker side of the plant and thus increases growth on this side by inducing cell elongation.

phreatic the term applied to volcanic activity, gases, etc, generated by contact between hot MAGMA and water derived from the ground, sea or lake. The proximity of the magma vapourizes the water, which builds up pressure, and ultimately this exceeds the pressure of the water holding in the gases, with the resultant eruption of steam and perhaps some magmatic material.

phyllite a low-grade metamorphic rock, pelitic in composition (*see* PELITIC ROCKS). It often has a shiny appearance due to the alignment of minerals such as MICA, chlorite and others, within the FOLIATION.

phylum (*plural* **phyla**) a part of the taxonomic classification of the animal kingdom. A phylum includes one or more classes that are closely related. Examples are Protozoa, Arthropoda, Chordata. In the classification of plants, the term division is used.

physical chemistry the study of the link between physical properties and chemical composition, and the physical changes caused by chemical reactions.

physics the study of matter and energy and changes in energy without chemical alteration. Physics includes the topics of magnetism, electricity, mechanics, heat, light and sound. The study of modern physics also encompasses quantum theory, atomic and nuclear physics. In combination with other disciplines, physics forms new topics, e.g. geophysics, biophysics.

phytoalexin compounds in plants that are instrumental in disease resistance. In the main phenolic (*see* PHENOL) or TERPENOID (*see also* TERPENES), they are produced or increased in concentration to restrict the growth of, or to destroy, fungi.

phytochemistry the study of the chemical make-up of plants.

pi (π) bond the COVALENT BOND formed when two atoms join to form a diatomic molecule. Pi bonds are discussed in terms of molecular ORBITALS, in which the shared electrons orbit the whole molecule rather than an atom. Pi bonds hold the molecule together by forming two regions of electron density above and below an axis between the bonded nuclei of the two atoms.

piedmont glacier an extension of ice from a valley glacier, which projects beyond its valley walls onto the adjacent flat plain at the foot of the mountains (i.e. the piedmont). Because the ice is now at a lower altitude, it may show more rapid diminution due to melting.

piedmont gravels coarse deposits of pebbles, BRECCIA, etc, found on the flat lowlands (piedmont) deposited by fast-flowing mountain rivers with a high sediment load. Upon reaching the flatter ground, the river slows and cannot carry as great a load, hence the larger material is deposited.

piezoelectric effect an effect of certain ANISOTROPIC crystals whereby opposite charges are generated on opposite crystal faces by the application of pressure. QUARTZ is such a crystal. One use of this phenomenon is in the crystal microphone.

pillow lava subaqueous, basaltic LAVA flows are characterized by pillow structures, each rarely more than one metre in diameter but often forming a sequence hundreds of metres thick. Pillow lavas form from long flow tubes, and as the sea water causes the rapid cooling of each lobe, so further branching and budding occurs to continue the outpouring.

pilot balloon a small hydrogen-filled balloon used in meteorology to determine wind speed and direction at high altitude.

pipette originally a glass tube with a fine tip and sometimes a bulbous central portion, used for obtaining a given volume of solution. It is filled by sucking or applying negative pressure by means of a rubber bulb. Automatic pipettes are now more commonplace.

pisolite a limestone made up of pisoliths, which are similar to ooliths (*see* OOLITE) but larger (> 2 mm). Pisoliths also show concentric internal layering.

pituitary gland *see* **endocrine system.**

place value notation when a number has more than one digit, the position, or place value, of each digit in the number is used to indicate what it is worth. For example, using the decimal system, the 6 in 362 and in 3620 stands for something different: in 362 it means 6 "tens" and in 3620 it means 6 "hundreds."

placers *or* **placer deposits** deposits rich in mineral ores, e.g. platinum, gold, cassiterite (tin oxide). They are produced by the mechanical action of weathering, creating a concentration of these minerals while lighter ones have been removed.

Planck's constant the proportionality constant (h) used in the equation to define the energy of a QUANTUM. Planck's constant has a value of 6.6262 x 10^{-34}Js and is named after the German mathematician and physicist Max Planck (1858-1947), who proposed the theory that radiant energy consisted of quanta.

planet the name given originally to seven heavenly bodies that were thought to move among the stars, which were themselves stationary. The term now applies to those moving in definite orbits about the SUN, which, in order of distance, are: MERCURY, VENUS, EARTH, MARS, JUPITER, SATURN, URANUS, NEPTUNE and PLUTO. Mercury and Venus are termed the inferior planets and Mars to Pluto the superior planets, the latter because they revolve outside the Earth's orbit.

planetary nebula a layer of glowing gas produced by and surrounding an evolved star (*see* HERTZSPRUNG-RUSSELL DIAGRAM) and representing a late stage in star evolution. (It is nothing to do with a planet.)

planetoid *see* **asteroid**.

plankton very small organisms, of plant and animal origins, that drift in water. The plants (or phytoplankton) are mainly diatoms (unicellular ALGAE) which PHOTOSYNTHESIZE and form the basis of the food chains. The animals or zooplankton feed on the diatoms and include small crustaceans and the larval stages of larger organisms.

plasma in biology, the same as BLOOD PLASMA. In physics, essentially a high temperature gas of charged particles (ELECTRONS and IONS) rather than neutral atoms or molecules. A plasma is electrically neutral overall, but the presence of charged particles means that it can support an electric CURRENT. It is of significance to the study of controlled NUCLEAR FUSION.

plasmid a DNA structure that exits outside the

chromosome and is able to replicate independently. Bacterial plasmids are used to produce recombinant DNA for gene cloning.

plasmolysis *see* **osmosis**.

plastics a group name for mainly synthetic organic compounds, which are mostly POLYMERS (formed by polymerization) that, when subjected to heat and pressure, become plastic and can be moulded. There are two types: thermosetting and thermoplastic materials. Thermosetting plastics are materials that lose their plasticity after being subjected to heat and/or pressure. Thermoplastic materials become plastic when heated and can be heated repeatedly without changing their properties.

plate (*see also* PLATE TECTONICS) the concept of plates arose from the observation that large areas of the Earth's crust have suffered little distortion and yet have travelled many kilometres. Thus plates have little seismic or volcanic activity but are fringed by margins that exhibit earthquakes, volcanism and mountain chains (whether submarine or subaerial). There are six major lithosphere plates: Eurasian, American, African, Indo-Australian, Antarctic and Pacific, and numerous smaller plates (e.g. Caribbean and Philippine) and microplates (e.g. the Hellenic in the eastern Mediterranean).

plateau eruptions *and* **plateau basalts** volcanic eruptions that form extensive lava flows, infilling topographic lows and accumulating to, ultimately, form a plateau. The lavas comprise numerous individual flows from several "shield" volcanoes, i.e. those characterized by fluid lava, little explosion, and a low angle cone.

platelet *see* **blood**.

plate tectonics *see* **continental drift**.

platinum metals a block of six TRANSITION ELEMENTS with similar properties—ruthenium (Ru), rhodium (Rh), palladium (Pd), osmium (Os), iridium (Ir) and platinum (Pt). The platinum metals are commonly found together, with gold and silver. Of the group, platinum is of greatest importance. It is very stable and is used mainly for jewellery, special scientific equipment, and chemical electrodes. It is also used as a CATALYST (*see also* ZEOLITES).

Pleiades, The the open cluster in the Taurus constellation. The seven main stars each individually named form the well-known group.

pleochroism the property displayed by some minerals when viewed by transmitted light under the petrological microscope. Different colours are seen when viewed in different directions. The phenomenon is seen when a mineral is rotated in polarized light and is due to unequal absorption of light vibrating in different planes. BIOTITE mica is a good example, showing varying shades of brown.

plug *or* **neck** a cylindrical remnant of a volcano formed by the solidification of magma within the main feeder of the volcano. Plugs are commonly roughly circular, steep-sided and vary in size from a few metres to over a kilometre. In addition to magma, they may contain PYROCLASTICS or AGGLOMERATE The whole feature is produced by the erosion and removal of the rocks surrounding the plug. Also known as **puy**, after Le Puy in the Auvergne region of France.

plume a geological plume is partly molten mantle material that ascends and that is thought to be responsible for volcanic activity within plates. In meteorology, a plume refers to snow that is blown over the ridge of a mountain. Nearer to its original meaning, in biology, plume refers to a feather or any structure resembling a feather.

plunge in general, the angle between a line and a horizontal datum. It is much used in FOLD terminology, where the fold plunge is the attitude of the fold axis. The angle (between the axis and the datum) is measured in a vertical plane and enables a full orientation to be given, e.g. a fold plunges 25° towards 135° (i.e. towards that bearing).

Pluto (*see* APPENDIX 6 for physical data) the ninth and smallest planet of the SOLAR SYSTEM, discovered in 1930. Little is known about the planet, but in 1979 its satellite Charon was discovered. Charon has a relatively large diameter compared to Pluto itself (600 : 2500 km), and in effect Pluto and Charon are a double minor-planet system.

pneumatolysis the late-stage magmatic process whereby hot volatile solutions cause changes in rock chemistry and mineralogy. The solution is released from the inner parts of the crystallizing body, and the volatiles (H, F, Cl, B) are carried through fissures, altering the rocks in the upper zone of the body.

podsol a soil with minerals leached from its surface layers into lower layers. Podsolization is an advanced stage of leaching, which involves the removal of iron and aluminium compounds, humus and clay minerals from the topmost horizons and their redeposition lower down.

point of inflection a point where a plane curve changes from the concave to the convex, relative to some fixed line, i.e. the point where it "crosses its tangent." At this point, the second DERIVATIVE of the function determining the curve is zero, i.e.

$$d^2y/dx^2 = 0.$$

Poiseuille's formula an expression that examines the relationship between the volume flow rate (Q) of a pipe and the radius of the pipe (a), the pressure gradient along the pipe (p/l), and the viscosity of the liquid carried by the pipe (η). Poiseuille's formula for the average volume per second ($m3s^{-1}$) is as follows:

$$Q = \frac{\pi pa^4}{8\eta l}$$

Poiseuille's formula does not apply to turbulent motion in pipes, but it is useful for examining pipes in which flow patterns have developed and can even be used to investigate blood flow in arteries or veins or water flow in plant stems or roots.

Poisson's ratio the ratio of the lateral strain to the longitudinal strain in a stretched piece of a material.

polar angle the angle between the positive (polar) axis and the radius vector in POLAR CO-ORDINATES.

polar axis the diameter of a sphere that intersects both poles. Also, in an equatorial telescope, it is the axis (parallel to the Earth's) the telescope revolves around so that the object is kept in the field of view.

polar co-ordinates the position of a point in space as represented by the co-ordinates (r, q), where q is the angle between the positive x-axis and a line from the origin to the point, and r the length of that line.

polar covalent bond the joining of two ATOMS due to the strong but unequal sharing of their ELECTRONS, which gives the bond, and thus the molecule formed, partial charges. Polar covalent bonds are formed between atoms that differ in their ability to attract electrons, with one atom having a greater ELECTRONEGATIVITY than another. For example, a hydrogen chloride molecule (HCl) has a polar covalent bond, as the chlorine atom is more electronegative than the hydrogen atom. The net result is that the chlorine end of the bond has a denser electron cloud because of its greater attraction for electrons. This causes the chlorine end of the molecule to have a partial negative charge, and as the whole HCl molecule is neutral, the hydrogen end has a corresponding partial positive charge.

polarimetry the measurement of OPTICAL ACTIVITY in, for example, sugar solutions. Also, the measurement of the extent to which light reflected from planetary surfaces is polarized.

Polaris the brightest star in the Ursa Minor constellation, once much used for navigation (in the northern hemisphere).

polarization the process by which the particles of a light wave are made to vibrate in one particular plane rather than the many directions taken by particles of normal light. Only transverse waves, and not longitudinal waves, can be polarized, so all electromagnetic waves can be polarized but longitudinal waves, such as sound, cannot. Some natural crystals, e.g. quartz and calcite, can polarize light because of their internal structure, and polarization has many scientific uses.

polarizing monochromator a device for producing a narrow band of light, used in studying the solar CHROMOSPHERE. It comprises a filter made up of QUARTZ crystals and CALCITE (or polaroid).

polar wandering the path created by the palaeo-magnetic pole over time. In concert with and after the various theories of CONTINENTAL DRIFT, palaeo-magnetic reconstructions of past periods showed a common magnetic pole in the Triassic when the continents were together in PANGAEA. These poles are now scattered, and the construction of the apparent polar wander curves indicates the tracks of magnetic north pole for the various continents with time. The sudden change in the path could be related to plate tectonic events such as collisions.

pollen analysis (*see also* PALYNOLOGY) a useful tool in reassembling the history of the flora of an area. Because the outer layer of pollen grains is resistant, particularly if deposited under the anaerobic conditions of rapid sedimentation, or in peat or stagnant water, they are widely distributed and preserved. Such analysis contributes to studies of climate change and sediment dating.

polychloroethene *see* **PVC**.

polygon a closed plane figure with three or more straight line sides. Common polygons are figures such as the triangle, quadrilateral and pentagon. A square is an example of a regular polygon, one in which all sides and all angles are equal. A general equation exists for the sum of the interior angles of a polygon with n sides:

sum of the interior angles = $180° (n - 2)$.

polyhedron a solid figure composed of four or more polygonal plane faces. The more faces it has, the closer it is to a sphere. A cube is an example of a regular polyhedron as all the faces and all the angles of a cube are equal.

polymer a large, usually linear MOLECULE that is

formed from many simple molecules, MONOMERS. Natural polymers include starch, cellulose (found in the cell walls of plants), and PROTEINS. Many synthesized polymers, such as nylon and polythene, are formed from ALKENES by ADDITION POLYMERIZATION.

polynomial in mathematics, an algebraic expression consisting of three or more terms, each of which is the product of a constant and one or more variables raised to a positive or zero integrated power. In biology, polynomial denotes a species name of more than two terms.

polypeptide a single, linear MOLECULE that is formed from many AMINO ACIDS joined by PEPTIDE BONDS. Polypeptides differ greatly in the number of amino acids they contain (usually from 30 to 1000). Although there are only 20 different amino acids, there are a huge number of possible arrangements in a polypeptide or PROTEIN, as the amino acids can be in any order. Most proteins consist of more than one polypeptide rather than a single polypeptide chain.

polytetrafluoroethene *see* **PTFE**.

polyvinyl acetate, polyvinyl chloride *see* **PVA, PVC**.

population inversion an essential process for LASERS (and MASERS), in which the system contains a higher number of particles with high energy than low—the reverse of the usual situation as specified by the Boltzmann Principle (a statistical distribution of large numbers of small particles, thermally agitated and acted upon by magnetic, electrical or gravitational fields). This inversion is achieved in the laser by input of energy from an external light source.

population types the two categories of star groups. Population I stars include hot blue stars found in the arms of spiral galaxies. Red stars, which occur in globular clusters, and the central parts of galaxies constitute Population II.

porosity the total of all the spaces in a rock is called the *absolute* porosity. However, not all these spaces will be connected to each other to permit the passage of fluids. The proportion of the rock that does contain interconnected pores is called the *effective* porosity, and it is stated as a percentage of the bulk volume of the rock. The way in which the pores are connected is called the *tortuosity*.

porphyrins naturally occurring pigments that include CHLOROPHYLL and the haem part of HAEMO-GLOBIN.

position angle the means of measuring the position of one point, in relation to another, on the celestial sphere.

positron a particle with the same mass as the ELECTRON but a positive electrical charge. Positrons are produced during decay processes (*see* BETA DE-CAY) and themselves are annihilated on passing through matter (*see* ANTIMATTER).

potassium (K) a soft alkali metal that is highly reactive. Combined with other elements it occurs widely, in silicate rocks as alkali FELDSPAR, in blood and milk, and also in plants. Potassium is used primarily and extensively in its compounds as fertilizers; potassium hydroxide is used in batteries and ceramics, and alloyed with SODIUM it can be used as a heat-transfer medium as a coolant.

potential difference the work done in driving a unit of electric charge (one COULOMB) from one point to another in a current-carrying CIRCUIT. The unit of potential difference is the VOLT, and the potential difference is frequently referred to as the VOLTAGE.

potential energy any energy stored within a body that can be used to do work. A body is said to have potential energy (U) when it has been raised from a resting point A against gravitation to resting point B. The potential energy of such a body can be derived from U = mgh, where m is the mass of the body, g is 9.8 ms^{-2} (GRAVITY), and h is the distance moved. If the body is released from resting point B, its potential energy is transformed into the energy of motion, KINETIC ENERGY. Potential energy is present in a spring that has been stretched and can also be stored in the form of chemical or electrical energy.

potential evapotranspiration the maximum quantity of water vapour that, in theory, can be released into the atmosphere by both evaporation and transpiration from an area of green vegetation that is fully supplied with water.

potential temperature the temperature of a sample of air when brought adiabatically (*see* ADIABATIC CHANGE) to a standard pressure.

power the rate at which work is done by or against a FORCE. Power is also regarded as the rate at which energy is converted. The unit of power is the WATT (W), which is equal to the transfer of one energy joule per second, i.e. 1W = 1Js^{-1}. Electrical power can be calculated by multiplying the voltage by the current, P = IV.

power notation the use of a small number (an EXPONENT) placed next to an ordinary number to show how many times the ordinary number is multiplied by itself e.g.:

3^5 ("3 to the power 5") means $3 \times 3 \times 3 \times 3 \times 3$

power series a series of functions of the form:

(A) $a_0 + a_1x + a_2x_2 + ... + a_nx_n + ...$

or

(B) $a_0 + a_1(x - a) + a_2(x - a)^2 + ... + a_n(x - a)^n + ...$

where x is a real VARIABLE and a represents constant COEFFICIENTS.

pozzolana a volcanic soil deposit that, when mixed with lime, produces a cement. It was first used by the Romans when the material was discovered near Pozzuoli, close to Naples. Pozzolanas are formed only when volcanic activity has been explosive, producing a vitreous PYROCLASTIC material. The term is now used collectively for all materials that exhibit reactivity with lime and that set in the presence of water.

precessional motion a rotating body precesses when the application of a couple (two equal and opposite parallel forces) with an axis at 90° to the rotation axis causes the body to turn around the third common perpendicular axis. A gyroscope exhibits this behaviour.

precession of the equinoxes a movement to the west of the equinoxes (*see* EQUINOX) due to the PRECESSIONAL MOTION of the Earth. This is caused by the gravitational attraction of the Sun and Moon on the equatorial bulge of the Earth.

precipitation the formation of an insoluble substance (precipitate) during a reaction between two solutions. Precipitation occurs because the IONS of the two substances involved exchange partners. For example, if a solution of silver nitrate, $AgNO_3(aq)$, is mixed with a solution of sodium chloride, $NaCl(aq)$, the insoluble silver chloride $AgCl(s)$ forms a precipitate. From the chemical equation for this reaction:

$AgNO_3(aq) + NaCl (aq) \longrightarrow AgCl (s) + NaNO_3 (aq)$

it can be seen that the chloride ion (Cl^-) has displaced nitrate.

In meteorology, rain, hail or snow falling from clouds onto the surface of the Earth.

pressure the FORCE exerted per unit area of a surface. The pressure of a gas is equal to the force that its molecules exert on the walls of the containing vessel, divided by the surface area of the vessel. The pressure of a gas will vary with its temperature and volume, as stated by BOYLE'S LAW, CHARLES' LAW, and the GAS LAWS. At any depth, the pressure in a liquid or in air equals the weight above the unit area, and therefore as the depth increases, the pressure also increases. This is also the reason for air pressure decreasing as height above sea level increases. The unit of pressure is the PASCAL, although air pressure is commonly measured using mercury BAROMETERS and hence has units of millimetres of mercury (mm Hg) corresponding to the varying mercury levels as air pressure changes.

pressure gradient the rate of change of (atmospheric) pressure on the ground as shown by ISOBARS.

pressure law *see* **gas laws**.

prime number a number that can only be divided by itself and 1. The first ten prime numbers are 2, 3, 7, 11, 13, 17, 19, 23, 29.

prism in mathematics, a solid with equal and parallel POLYGONS as ends and parallelograms as sides.

In physics, a prism is triangular in shape and made of transparent material, and used to deviate or disperse a ray in optical instruments or laboratory experiments.

procaryote any organism that lacks a true-membrane NUCLEUS and is either a bacterium or a blue-green algae (cyanobacteria). Procaryotes have a single CHROMOSOME and do not undergo MEIOSIS or MITOSIS as they lack the MICROTUBULES to form the spindle. Procaryotes replicate by a form of asexual reproduction, called binary fission, in which the two sister chromosomes are attached to separate regions on the cell membrane, which starts to fold to form a cleavage. The cell eventually forms two daughter cells after the CYTOPLASM has been completely split by the fusion of the enfolding cell membrane.

product rule a method used in CALCULUS to differentiate the product of two functions. If there are two functions, u and v, then the product function f(x) = uv can be differentiated using the following formula:

$$f(x) = u'v + uv'$$

progesterone a STEROID hormone, which in mammals is important in pregnancy.

prognostic chart a weather forecast chart specifying the expected meteorological conditions.

prominence a band of higher density and lower temperature glowing gas, seen in the CHROMOSPHERE and lower part of the CORONA. They are best seen at the edge of the Sun during an eclipse.

proper motion the part of a star's motion in space that is perpendicular to the line of sight and relative to the Sun or another star.

prophase the first stage of MEIOSIS or MITOSIS in cells of EUCARYOTES. During prophase, the CHROMOSOMES condense and can thus be studied using a microscope.

protease a group of ENZYMES that act as catalysts in the breaking up of PROTEINS into PEPTIDES and AMINO ACIDS. Examples are PEPSIN and trypsin.

protein a complex, nitrogen-containing, organic compound of vital significance to all living matter. Proteins have high MOLECULAR WEIGHTS and comprise hundreds or thousands of AMINO ACIDS joined to form POLYPEPTIDE chains. The amino acid sequence confers upon each protein its particular properties. ENZYMES are another important group of proteins.

proteolysis the splitting of proteins through the catalytic action of PROTEASES. To break down a protein completely (into its amino acids) usually requires several proteases acting one after the other.

proton a particle that carries a positive charge and is found in the NUCLEUS of every ATOM. As an atom is electrically neutral, the number of protons equals the number of negatively charged ELECTRONS. Although the MASS of a proton (1.673×10^{-27} kg) is far greater than the mass of an electron (9.11×10^{-31} kg), their charges are equal in magnitude. The number of protons in the nucleus of an atom (ATOMIC NUMBER) is identical for any one element and is used to classify elements in the PERIODIC TABLE. For example, as every oxygen atom contains 6 protons, it has an atomic number of 6 in the periodic table, whereas every gold atom contains 79 protons and thus has an atomic number and periodic table position of 79.

Proxima Centauri the star nearest to the Sun and a member of the constellation Centaurus, some 4.3 light years away.

psammite a rock rich in the mineral QUARTZ, and formed by the metamorphism of a SANDSTONE, QUARTZITE or ARKOSE.

pseudopodium (*plural* **pseudopodia**) the temporary projection from the body of certain cells. Pseudopodia are formed in simple, single-celled organisms, such as amoeba, as a mechanism for locomotion and food intake. They are also formed by white blood cells, which use PHAGOCYTOSIS to ingest particles.

PTFE (*abbreviation for* **polytetrafluoroethene**) a thermosetting PLASTIC produced by the polymerization of tetrafluoroethene (CF_2CF_2). Under its trade names of Teflon and Fluon, it is used to line saucepans, where its chemical unreactivity and heat resistance are useful. It is also used in engineering applications.

pulmonary artery one of the two arteries that carry deoxygenated blood from the HEART to the lungs, where it is oxygenated. The pulmonary arteries are the only ones that carry blood with a high concentration of carbon dioxide rather than a high concentration of oxygen. All other arteries carry oxygenated blood to the tissues, where oxygen is exchanged for carbon dioxide.

pulmonary vein one of the four veins that carry oxygenated blood from the lungs (two veins leave both the left and right lungs) to the left ATRIUM of the HEART. The pulmonary veins are unique, as they carry oxygenated blood while all other veins carry deoxygenated blood back to the heart after it has exchanged oxygen for carbon dioxide in the various tissues of the body.

pulsar a star that is a sorce of radio frequency radiation (*see* ELECTROMAGNETIC WAVES) which is emitted in regular short bursts. Many have been located with radio telescopes, and it is thought that they are collapsed, rotating NEUTRON STARS.

pumice an acidic rock, usually of PYROCLASTIC origin, that occurs as a vesicular frothy glass formed by ejection from a volcano followed by rapid cooling.

purine one of the two different structures that form the base components of DNA and RNA. A purine has a double ring structure that consists of both carbon and nitrogen atoms. The bases, ADENINE and GUANINE, are both purines that will form hydrogen bonds with their complementary PYRIMIDINE bases to form the double helix of the DNA molecule.

putrefaction the breakdown (decomposition) of plants and animals after death by anaerobic bacteria.

puy *see* **plug**.

PVA (*abbreviation for* **polyvinyl acetate**) a PLASTIC produced by the polymerization of vinyl acetate. It is used in coatings, adhesives and inks.

PVC (*abbreviation for* **polyvinyl chloride** *or* **poly-chloroethene**) the most widely used of the vinyl PLASTICS formed by POLYMERIZATION of vinyl chloride (chloroethene H_2CCHCl). PVC is used for pipes, ducts, mouldings and as a fabric in clothing and furnishings.

pyranometer a device for measuring global solar radiation. In physics, the instrument comprises two metallic strips of different thicknesses, which are black to absorb the heat. The difference in thickness means each strip reaches a different temperature.

pyrgeometer an instrument for measuring the loss of heat from the Earth's surface by radiation. It consists of a number of polished and blackened surfaces that cool at different rates.

pyrheliometer a device for the measurement of direct solar radiation.

pyrimidine one of the two different structures that form the base components of DNA and RNA. A pyrimidine has a single ring, consisting of both carbon and nitrogen atoms. The bases, cytosine, THYMINE and URACIL, are all pyrimidines.

pyrite otherwise known as fool's gold, FeS_2. The sulphide mineral occurs widely distributed in numerous environments. It forms an accessory mineral in IGNEOUS ROCKS and SEDIMENTARY ROCKS, particularly black shales (which were deposited in anaerobic conditions). It also occurs in replacement deposits and some metamorphic rocks. It was mined for the production of sulphuric acid, but no longer.

pyroclastic rocks rocks formed by the violent expulsion of rock and lava from volcanic vents. This may include PUMICE, IGNIMBRITES, TEPHRA and AGGLOMERATES.

pyroxenes an important group of rock-forming minerals that are silicates of iron, magnesium and calcium, sometimes with aluminium. Certain varieties contain sodium or lithium. All are characterized by the Si_2O_6 chain structure, and there are many varieties due to replacement of one metal by another. The group can be divided into two on their crystal systems, but both show very good CLEAVAGES. Pyroxenes are widely distributed in igneous and metamorphic rocks and show a variety of colours, but usually dark greens, brown or black. Common forms are diopside, enstatite and augite, the latter being dominant and occurring in BASALT, GABBRO, DOLERITE and many other rocks.

pyruvate a colourless liquid formed as a key intermediate in the metabolic process of GLYCOLYSIS and the production of ATP.

Pythagoras' theorem the geometrical theorem that states that in any right-angled triangle, the square of the HYPOTENUSE is equal to the sum of the squares of the two shorter sides. This theorem is named after the Greek philosopher and mathematician of the 4th century BC. For a given right-angled triangle in which the sides are x and y units long, the hypotenuse (h) can be obtained from $h^2 = x^2 + y^2$. Pythagoras' theorem provides a method of calculating the length of any side of a right-angled triangle if the lengths of the other two sides are known.

quadratic equation an equation that has the general form $ax^2 + bx + c = 0$. The roots of any quadratic equation can be obtained from the formula:

$$x = \frac{-b \pm \sqrt{b^2 - 4ac}}{2a}$$

Part of this equation, $b^2 - 4ac$, is called the discriminant and describes the roots of a quadratic equation. If the discriminant has a positive value ($b^2 - 4ac > 0$) then the roots are real and distinct, but if the discriminant has a negative value ($b^2 - 4ac < 0$), then the roots are imaginary (graph does not cut x-axis). If the discriminant is zero ($b^2 - 4ac = 0$), then the roots are real but not distinct as both roots have the same value.

quadrature the position of the Moon or a superior planet such that a line joining it to earth is at right angles to a line joining the EARTH to the SUN.

quadrilateral any geometric shape that has four sides. Some examples of quadrilaterals are the rhombus, kite, parallelogram and the rectangle.

qualitative analysis the chemical analysis of a sample to identify one or more constituents.

quantitative analysis determination of the relative amount of species making up a sample, which usually refers to elemental analysis.

quantum (*plural* **quanta**) a small, discrete quantity of radiant energy. Electromagnetic radiation (*see* ELECTROMAGNETIC WAVES) is explained in terms of small particles as well as waves, as it is assumed that it can be absorbed or emitted in quanta. The energy of one quantum (E) is derived from the equation, $E = hv$, where v is the FREQUENCY of the radiation and it is PLANCK'S CONSTANT.

quantum numbers a set of four numbers used to describe atomic structures (*see* ORBITALS). The first, n, the principal quantum number, defines the shells (stationary orbits in BOHR'S THEORY) which are visualized as orbitals. The orbit nearest the NUCLEUS has $n = 1$, and contains 2 ELECTRONS. The second shell, $n = 2$, contains 8 electrons, the maximum number of electrons in each shell being limited by the formula $2n^2$; the orbital quantum number, l, defines the shape of the orbits within one shell, which are designated s, p, d, f orbits; the magnetic orbital quantum number, m, which sets the spatial position of the orbit within a strong magnetic field; and s the spin quantum number, based upon the assumption that no two electrons may be exactly alike, and thus opposite spins are invoked for pairs of electrons.

quark any of the theoretical building blocks that participate in the strong interactive forces between ELEMENTARY PARTICLES. Originally, it was postulated that there were three types of quark, each carrying a charge that is less than that of one electron, but the need to explain new phenomena may make it necessary to have more types of quarks.

quarter the phase of the Moon at QUADRATURE, i.e. first quarter and last quarter. The other two quarters are full and new moon.

quartering a method used in mineral extraction (and geochemistry) whereby to obtain a representative sample for analysis a cone of material is divided into four. Two opposite quarters are rejected and another cone formed from the remainder. The process is then repeated until a sample of the required size remains.

quartz one of the most widely distributed ROCK-forming minerals, SiO_2. It occurs in all kinds of rocks, and in its various crystalline forms and with certain impurities, it forms semi-precious stones, e.g. amethyst and agate. Quartz crystals exhibit the PIEZOELECTRIC EFFECT.

quartz wedge a so-called accessory plate used in optical mineralogy to help in mineral identification (mica and gypsum are also used). The wedge is inserted into the microscope to estimate the BIREFRINGENCE and to determine the optical sign of certain minerals.

quartzite a rock formed by the metamorphosis of quartz SANDSTONES and thus itself composed mainly of quartz. Deformation can produce oriented, elongate grains, which may create a FABRIC in the rock. SEDIMENTARY quartzites are sandstones with quartz as the cement, and to distinguish them from metamorphic or metaquartzites, are termed orthoquartzites.

quasar any of the "quasi-stellar" objects, which are extremely compact, light-emitting and yet enormously distant bodies—up to 10^{10} LIGHT YEARS away.

quenching the rapid cooling of magma, which in addition to producing glassy rocks when it occurs naturally, is also used in experimental petrology. Because the magmatic reactions are halted quickly, quenching provides an indication of the mineral phases in equilibrium under the existing temperature and pressure.

quinine a colourless ALKALOID with a very bitter taste, which was used in the treatment of malaria.

quotient the result obtained when a mathematical quantity (number, function or equation) is divided by another quantity.

quotient rule a mathematical method used in CALCULUS to differentiate the QUOTIENT of two functions. The quotient rule for the functions u and v, $f(x) = u/v$, is as follows:

$$f'(x) = \frac{u'v + uv'}{v^2}$$

327

R

racemic isomers mixtures of STEREOISOMERS that are optically inactive. The optically active components can be obtained by various chemical means.

racemization the production of a racemic, inactive substance from the optically active form.

radar (*acronym for* Radio Detection And Ranging) the use of radio waves to detect the presence and distance of objects. Used in navigation of aircraft, ships, missiles and SATELLITES.

radar astronomy the use of radio waves within the SOLAR SYSTEM for purposes of distance measurement and surface mapping (e.g. of VENUS).

radian an alternative to the degree in measuring angles, derived from the angle subtended at the centre of a circle by an arc equal to the length of the radius. There are 2π (6.284) radians in a full circle (360°).

radial symmetry the structural arrangement of an organism (or organ) so that a plane bisecting the structure, in any direction, gives equal and opposite halves—mirror images. Obvious examples include the coelenterates (jellyfish) and echinoderms (starfish and sea urchins), and also plant stems and roots. When applying this concept to flowers, the term actinomorphy is used.

radiant a point on the CELESTIAL SPHERE that is the

origin for parallel tracks through space, e.g. as shown by meteors in a shower.

radiation the emission of energy from a source, applied to ELECTROMAGNETIC WAVES (radio, light, X-rays, infrared, etc), particles (α, β, protons, etc), and sound.

radiative equilibrium an idealized situation for matter in stars where the temperature everywhere creates a gas pressure to balance the gravity of the star.

radical a group of atoms (within a compound), usually unable to exist independently, which is unchanged in reactions affecting the rest of the molecule. (Now often referred to as a group.)

radical sign the inverse operation to forming a POWER is that of extracting a root and is expressed by a radical sign $\sqrt{}$, e.g. $\sqrt{4} = 2$

radioactivity the emission of α or β particles and/or γ rays by unstable elements, while undergoing spontaneous disintegration.

radio astronomy the detection of a large range of radio waves emitted by numerous sources, including the SUN, PULSARS, remnants of SUPERNOVAe, and QUASARS.

radiobiology the study of biological systems as influenced and affected by radioactive materials, and the use of carefully controlled amounts of radioactive substances (tracers) to study metabolic processes.

radiocarbon dating a method of dating organic material, although it is only applicable to the last 6000-8000 years. There is a small proportion of radioactive ^{14}C in the atmosphere, which is taken up naturally by plants and animals. When an organism dies, the uptake ceases and the ^{14}C decays with a HALF-LIFE of 5730 years. Comparison of residual radioactivity with modern standards enables an age to be calculated for a sample.

radiochemistry the study of the science and techniques for purification of radioactive (*see* RADIOACTIVITY) materials (ISOTOPES and their compounds).

radiography the process of producing an image of an object on photographic film (or on a fluorescent screen), using X-rays (or a similar short wavelength electromagnetic wave, e.g. gamma rays). The photograph thus produced is termed a radiograph, and the process is used widely in diagnostic medicine.

radiolarian ooze a deep-sea ooze that contains a significant proportion of radiolarian tests (protective "shells" made of silica).

radiometric dating a precise method of rock dating that relies upon the relative amounts of a radioactive element present in original and decay states, i.e. the so-called "parent" and "daughter" isotopes. URANIUM-lead was an early system, but there are now more, including potassium-argon (K-Ar) and rubidium-strontium (Rb-Sr). Uranium-238 decays to lead-206 with a HALF-LIFE of 4.5 billion years; rubidium-87 decays to strontium-87 (half-life 50 billion years); and potassium-40 decays to argon-40, with a half-life of 1.5 billion years. There are others, including samarium-neodymium (Sm-Nd) and thorium-lead (Th-Pb). The Rb-Sr method is used for dating granitic rocks (*see* GRANITE) because rubidium mirrors potassium in minerals such as MICAS. More recent dates are determined using the K-Ar method.

radio (waves) ELECTROMAGNETIC radiation used to communicate through space without an intermediate physical link. The information thus conveyed can include sound, pictures, and digital data.

radiolysis the chemical decomposition of a substance subjected to ionizing radiation.

radionuclide any ISOTOPE of an element that undergoes natural radioactive decay.

radiosonde an instrument that is used in meteorology and that is carried through successive atmospheric levels by a balloon. The apparatus measures temperature, HUMIDITY and pressure, and the results are transmitted to a radio receiver.

radio telescope the instrument used to detect and analyse extra-terrestrial electromagnetic radiations. There are two types—the parabolic reflector, which focuses the radiation onto an aerial, and the interferometer, where an interference pattern of "fringes" is formed to enable precise wavelength measurement to be made. The latter is more accurate while the former is easily moved.

radius vector a line joining the focus of an orbit to any body moving about the focus in that orbit, e.g. the line from the Sun to a planet.

rain the condensation of water vapour into droplets when moist air is cooled below its DEWPOINT. The average diameter of a raindrop is 2 mm or less.

rainbow the characteristic display of colours formed by the REFRACTION and internal REFLECTION of sunlight by raindrops in the air.

rain gauge an apparatus for measuring rainfall, comprising a standard funnel leading into a collecting bottle.

rainmaking the attempt to produce rainfall by "seeding" supercooled clouds with, for example, solid carbon dioxide.

rain shadow the production of dry, or even desert, conditions on the landward side of mountains because most of the moisture from the winds blowing off the ocean or sea falls on the slopes facing the ocean. This occurs in the U.S.A., where the desert areas of Nevada and eastern California contrast with the wet, western side of the Coast Range and Sierra Nevada.

raised beach a beach that is now above the level of the shoreline. This may be due to earth movements or a fall in sea level.

RAM (*abbreviation for* random access memory) computer memory that can be written to, read from, altered and erased; a temporary working memory.

rank of coal coal occurs in different "ranks," starting with peat, then through the bituminous coals, to LIGNITE and anthracite. The increase in rank is due to a decrease in VOLATILES and an increase in carbon content. The volatiles include methane, hydrogen and carbon dioxide. Rank increases with burial depth (due to a temperature increase) and with time. Tectonic deformation and thermal metamorphism can also play a part, e.g. anthracites often are associated with areas of folding.

Raoult's law a law that states that the vapour pressure of an (ideal) solution is the sum of the vapour pressures of each component.

rare earth elements *see* **lanthanides**.

rate of change (*also called* the DERIVATIVE) is the slope of a graph $y = f(x)$ at a given point c, or in more precise terms, it is the limit, as h approaches zero, of $f(c + h) - f(c)/h$.

ratio numbers these are used to compare the sizes of two or more quantities. If in a class of 24 pupils there are 8 boys and 16 girls, the ratio of boys to girls is:

$$8 : 16 \text{ or } 1 : 2$$

It is usual to try to reduce one of the numbers to 1, and it is essential that the two quantities be expressed in the same units.

rational number a number that can be obtained by dividing one quantity by another quantity: a/b with b = 0. This includes all whole numbers and most FRACTIONS.

rayon the term applied formerly to "artificial silk," but now to two manmade cellulose fibres, viscose and cellulose acetate rayon.

reaction rim the outer rim of a mineral that reacts with, for example, the MAGMA in which it is contained to form an altered shell of a different mineral.

reagent a chemical substance or solution that is used to produce a characteristic reaction in chemical analysis.

real number any RATIONAL or IRRATIONAL number. Real numbers exclude imaginary numbers and COMPLEX NUMBERS.

recessive allele a gene form that is not expressed and will therefore not affect the PHENOTYPE of the organism unless the organism is HOMOZYGOUS for the recessive allele. Although an organism that is HETEROZYGOUS for a recessive allele will possess this allele, the dominant form of the gene will be expressed, thus masking the presence of the recessive form.

reciprocal the inverse ("other way up") of a FRACTION; the reciprocal of a number A is $1/A$. For example, the reciprocal of $5/12$ is $12/5$, and the reciprocal of 6 is $1/6$.

recombinant DNA a new DNA sequence formed by the insertion of a foreign DNA fragment into another DNA molecule. Recombinant DNA is used extensively throughout GENETIC ENGINEERING, when bacteria are frequently used as hosts for the expression of recombinant DNA molecules and the subsequent coding for the desired protein. It is particularly useful for producing a significant quantity of a human PROTEIN, such as INSULIN.

recombination *see* **genetic recombination**.

rectangle *see* **parallelogram**.

rectifier a device that converts ALTERNATING CURRENT into direct current.

recumbent fold a FOLD in which the axial plane is sub-horizontal or horizontal.

recurring decimal when a number or set of numbers is, after a certain point, repeated indefinitely. The recurring figures are often signified by dots, for example, $0.\dot{3}$ is equivalent to $0.333333\ldots$ Also called repeating decimal.

red blood cell *see* **erythrocyte**.

red giant a star that has consumed about 10 per cent of its hydrogen and moves out of the MAIN SEQUENCE in the HERTZSPRUNG-RUSSELL DIAGRAM. Red giants then consume their hydrogen at a greater rate, eventually contracting and becoming WHITE DWARFS.

red shift the light observed from certain galaxies shows a displacement of spectral lines towards the red end of the spectrum—hence red shift. This is interpreted as being due to the DOPPLER EFFECT and signifies that the galaxies are receding into space.

redox potential a method for evaluating the REDUCTION or OXIDATION potential of a reactant. Redox potentials are arranged on an arbitrary scale, which uses the standard hydrogen electrode as the reference redox reaction by assigning it a potential of zero volts. The strongest REDUCING AGENTS, i.e. those most easily oxidized, are at the top of the scale while the strongest OXIDIZING AGENTS, i.e. those most easily reduced, are at the bottom of the scale.

redox reaction a chemical reaction in which both REDUCTION and OXIDATION are involved. If the overall REDOX POTENTIAL for such a reaction has a positive value, then it is a spontaneous and feasible reaction.

reducing agent any substance that will lose ELECTRONS during a chemical reaction. Reducing agents will readily cause the REDUCTION of other atoms, molecules or compounds, depending on the strength of the reducing agent and the reactivity of the other reactant. The strongest reducing agents are active alkali metals such as lithium (Li), potassium (K), barium (Ba), and calcium (Ca).

reduction any chemical reaction that is characterized by the loss of oxygen or the gain of ELECTRONS from one of the reactants. A molecule is also said to be reduced if it has gained a hydrogen atom. There is always simultaneous OXIDATION if reduction has occurred in any reaction.

reduction formulae equations that allow the TRIGONOMETRIC FUNCTION of any angle to be expressed as the trigonometric function of an ACUTE angle. Tables giving the values of trigonometric functions for angles at various intervals have been computed, so the problem of finding the trigonometric function of any angle is reduced to finding the trigonometric function of an angle between 0° and 90°, looking this up in the table, and then prefixing the proper sign.

reflecting telescope a telescope that uses a mirror to focus light rays. There are several versions, including the NEWTONIAN TELESCOPE, named after the English physicist Isaac Newton (1642–1727), who first realized its potential. The largest telescopes in the world are all of the reflecting variety.

reflection the property of certain surfaces whereby rays of light falling upon them are returned (reflected) in accordance with definite laws. The incoming, or incident, ray becomes the reflected ray.

reflux when a liquid is boiled in a flask and a condenser is attached so that vapour condenses and is returned to the flask. This keeps the liquid boiling but prevents loss by evaporation.

refracting telescope a telescope that uses lenses to focus light rays, first applied to astronomy by the Italian physicist Galileo Galilei (1564–1642).

refraction the bending of, most commonly, a ray of light, on travelling from one medium to another. The refraction occurs at the interface between the media and is caused by the light travelling at different velocities in different media. The incident ray becomes the refracted ray upon refraction (*see* REFRACTIVE INDEX).

refractive index (n) the ratio of the SINE of the angle of incidence (the angle betweeen the incident ray and the line drawn PERPENDICULAR to the surface at that point) to the sine of the angle of refraction, when light is refracted from a vacuum into the medium.

regeneration the repair or regrowth of bodily parts of an organism that have been damaged and subsequently lost. Regeneration is rare in higher, complex animals but is quite common in lower, simpler animals in which the extent of regeneration can range from limb regeneration in crustaceans to the regeneration of the whole organism from one segment, as in certain annelid worms. Regeneration is common in plants and occurs naturally, as in VEGE-TATIVE PROPAGATION, or can be induced to propagate plants of economic importance, such as the potato and tobacco plants. Complete regeneration of any plant is only possible if its vegetative cells have retained the full genetic potential (i.e. are TOTIPO-TENT), enabling them to replicate every part of the plant.

regional metamorphism metamorphism involving both pressure and temperature, and possible shear stress, due to converging plates. The extent of orogenic belts (*see* OROGENY and MOBILE BELTS) means the effects are on a *regional* scale. The metamorphism can occur as several events tied in with several tectonic events, and the former can occur before, during or after a tectonic episode. The rock FABRIC changes with metamorphic grade (increasing in grain size) and moves from slaty cleavage to phyllitic, schistose and eventually gneissose. As seen with BARROVIAN METAMORPHISM, specific assemblages of minerals are generated relating to pressure/temperature regimes, and a whole scale of metamorphic FACIES can be produced, as established by the Finnish petrologist Pentti Elias Eskola (1883–1964) in 1920. The facies include amphibolite—middle to high grade regional metamorphism with the diagnostic minerals hornblende, (an AMPHIBOLE) and plagioclase FELDSPAR; greenschist—moderate pressure and temperature metamorphism creating assemblages with chlorite, actinolite (an amphibole), albite (plagioclase FELDSPAR) with QUARTZ and epidote; granulite—high-temperature and high-pressure conditions as found at the base of continental crust, producing mineral assemblages including PYROXENE minerals and PLAGIOCLASE.

regolith a fine powdery covering on the MOON (and other planets and asteroids), created by meteoritic impact. On the Moon, the material is several metres thick. In geological terminology, regolith refers to the layer of unconsolidated and weathered material that lies over solid rock. It may comprise rock fragments, mineral grains and soil components, and in the humid tropics can reach enormous thicknesses—commonly tens of metres.

regression in mathematics, the connection between the expected value of a random VARIABLE and the values of one or more possibly related variables. In biology, the tendency to return to an average state from an extreme one.

rejuvenation the action of a river system, caused by uplift of an area, when it increases its rate of erosion and cuts down as if it were a younger stream.

relative atomic mass (*formerly called* **atomic weight**) the mass of atoms of an element given in atomic mass units (u), where $1u = 1.660 \times 10^{-27}$ kg.

relativity the theory derived by EINSTEIN that establishes the concept of a four-dimensional space-time continuum where there is no clear demarcation between three-dimensional space and independent time, hence space and time are considered to be bound together. The important results of the theory include the appreciation that the mass of a body is a function of its speed; the derivation of the mass-energy equation, $E = mc^2$ (where c = speed of light), and the relative nature of time itself, i.e. there is no absolute value or interval of time.

remanent magnetization the magnetization "locked into" a rock at its formation, due to the Earth's magnetic field, or imposed during some later event.

repeating decimal *see* **recurring decimal**.

replication the duplication of genetic material, generally before cell division.

reservoir rock a porous and permeable rock, which can hold oil, gas or water. Typical rock types are SANDSTONE, LIMESTONE or DOLOMITE. Almost two thirds of oil occurrences are in sandstones and related lithologies (e.g. GREYWACKES), with almost one third in carbonate rocks. For example, dune sands in the Permian, which now underly the southern North Sea, the Netherlands and north Germany, form the reservoir for gas from the Coal Measures beneath. The CAP ROCK in this case is HALITE, which flowed to fill any fractures.

residual deposits the production of rock waste (from clays to boulders) due to weathering, or the weathered material remaining after dissolution of soluble components. In both cases the processes occur *in situ*.

resins natural resins are organic compounds secreted by plants and animals e.g. rosin, derived from pine trees. Synthetic resin is the term now applied to any synthetic PLASTIC material produced by polymerization.

resistance (R) measured in OHMS and calculated as the potential difference between the ends of a CONDUCTOR, divided by the CURRENT flowing. Superconductors apart, materials resist the flow of current to varying degrees, and some of the electrical energy is thereby converted to heat.

resistivity the reciprocal of a material's conductivity, giving the resistance in terms of its dimensions.

resistor a component of electric CIRCUITS, used to provide a known RESISTANCE.

resolution in chemistry, the separation of an optically inactive compound or mixture into its optically active components (*see* OPTICAL ISOMERISM).

resonance the creation of vibrations in a system by the application of a periodic force, e.g. from another vibrating system. As the FREQUENCY of the applied force becomes nearer to the natural frequency of the system, the vibrations increase, to reach a maximum when the two frequencies are equal.

resorption the partial solution or fusion of a well-formed PHENOCRYST in a MAGMA. It is due to changes in magma temperature, pressure or composition. If the magma is erupted rapidly then the phenocryst, which will now exhibit poor shape, is set in a fine GROUNDMASS.

respiration the process by which living cells of an organism release energy by breaking complex organic compounds into simpler ones using enzymes. Respiration can occur in the presence of oxygen (AEROBIC RESPIRATION) or in its absence (ANAEROBIC RESPIRATION) and has an initial stage called GLYCOLYSIS, which is common to both forms of respiration. The term respiration is also used, although less frequently, for gaseous exchange (better known as breathing) in an organism, which involves the uptake of oxygen from, and the release of carbon dioxide to, its surrounding environment.

retrogressive metamorphism *or* **retrograde metamorphism** rocks will recrystallize in response to a lowering of grade. However, recrystallization will only be accompanied by retrograde reactions if water remains or is introduced into the system. The mineral reactions will then produce hydrated minerals—the opposite of the dehydration of "prograde" metamorphism.

retrovirus *see* **virus**.

reversed fault a fault in which the hanging wall moves up relative to the footwall (*see* FAULT). Alternatively, it is where the direction of dip is towards the upthrown side.

reversible reaction a chemical reaction that can proceed in either direction. The incomplete reaction results in a mixture of reactants and products, and the balance can be altered by a change in the controlling factors, whether pressure, temperature or concentration.

rheostat a RESISTOR of variable RESISTANCE.

rhesus *see* **blood grouping**.

rhombus *see* **parallelogram**.

rhythmic sedimentation a repeated sequence of rock units, which form a pattern (*see* CYCLOTHEM). *Cyclic* sedimentation is a repetition of sedimentary FACIES, and the cycles may be symmetrical or asymmetrical. The cyclic nature may be due to repeated tectonic activity or to advances (or retreats) of the sea.

ribonucleic acid *see* **RNA**.

ribosomal RNA (rRNA) one of the three major classes of RNA, which is transcribed from DNA in a structure of eucaryotic nuclei called the NUCLEOLUS. Along with many PROTEINS, ribosomal RNA forms the cellular structures called RIBOSOMES, which are found in both EUCARYOTIC and PROCARYOTIC cells.

ribosome the cellular structure that is the site of PROTEIN synthesis in all EUCARYOTIC and PROCARYOTIC cells. Ribosomes are composed of one large and one small sub-unit, which contain RIBOSOMAL RNA and associated proteins. Analysis of procaryotic and eucaryotic ribosomes indicates that they share the same evolutionary origins as their structure, and the RNA they contain (except a segment unique to eucaryotes) are virtually identical. Ribosomes assemble at one end of a MESSENGER RNA molecule and move along the molecule to build the POLYPEPTIDE chains of all proteins in a process called TRANSLATION.

Richter scale the scale, devised by the American seismologist Charles Francis Richter (1900–85), used to measure the intensity of earthquakes. It uses the AMPLITUDE of seismic waves, which depends on the depth of the earthquake focus. Recording stations register the waves, and for a shallow earthquake the magnitude is given by:

$$M = \log (a/t) + 1.66 \log\Delta + 3.3$$

where a is the maximum amplitude, t the period (the time between a repeat of the same wave form) and Δ is the angular distance between the focus and the station. A slightly modified version is used for deeper earthquakes. Earlier systems of intensity measured, e.g. that devised by Giuseppe Mercalli (1850–1914), the Italian seismologist, relied more upon the effects seen or felt by observers when the seismic waves reached them. Below, in brief, is the arbitrary scale from 1 to 12:

1. *Instrumental* - detected only by seismographs.
2. *Feeble* - noticed by sensitive people.
3. *Slight* - similar to a passing lorry.
4. *Moderate* - rocking of loose objects.
5. *Rather strong* - felt generally.

6.	*Strong*	-	trees sway; loose objects fall.
7.	*Very strong*	-	walls crack.
8.	*Destructive*	-	chimneys fall; masonry cracked.
9.	*Ruinous*	-	collapse of houses where ground starts to crack.
10.	*Disastrous*	-	buildings destroyed; ground badly cracked.
11.	*Very disastrous*	-	bridges and most buildings destroyed; landslides.
12.	*Catastrophic*	-	total destruction; ground moves in waves.

ridge *see* **mid-oceanic ridge**.

right angle an angle of 90°.

right ascension a coordinate used, with others, in specifying positions on the CELESTIAL SPHERE.

ring compound *see* **closed-chain**.

ring dyke intrusive igneous bodies with a ring-shaped cross-section and steeply dipping walls. Ring dykes range in width from metres to hundreds of metres, and their diameter is usually up to 20–25 kilometres (12.5–15.5 miles).

ripple marks the preservation of ripples in sandstones, which may exhibit small cross-laminations (*see* CROSS-BEDDING) reflecting current movement.

RNA (*abbreviation for* **ribonucleic acid**) a NUCLEIC ACID that exists in all living cells. It is made up of a single-stranded chain of alternating ribose and phosphate units, BASE PAIRED between ADENINE and THYMINE or CYTOSINE and URACIL.

rock an aggregate of minerals or organic matter. Rocks are classified into three types: IGNEOUS, SEDIMENTARY, and METAMORPHIC.

ROM (*abbreviation for* read only memory) the part of the memory in a computer that is fixed and can be read but not written to or altered.

röntgen (R) a radiological term defining the X-ray or gamma-ray dose producing ions carrying a specific charge.

röntgen equivalent man a unit of radiation dose that has now been replaced by the SIEVERT.

rubber a natural hydrocarbon POLYMER (polyisoprene) from the *Hevea brasiliensis* tree. Items made from rubber are produced by adding various agents followed by vulcanization (heating in the presence of sulphur). Synthetic rubbers are polymers (or copolymers) of simple molecules.

rudaceous a general term for SEDIMENTARY ROCKS with a grain size of 2 mm or more.

ruminant any mammal that has four compartments in the stomach to aid the digestion of large amounts of plant matter. Ruminants (order Artiodactyla) include cattle, sheep, deer and giraffes. In the first section of their stomach, the rumen, food is enveloped in a mucus and is partially digested by an ENZYME called cellulase, which is supplied by the billions of bacteria living in the rumen. After this, the food is regurgitated to the mouth, and, after chewing, it passes through a further two sections (water is removed) and eventually ends up in the true stomach, the abomasum, which contains the enzymes essential for complete digestion.

sabkha a flat, coastal belt situated between desert dunes on the landward side, and a lagoon and the sea on the other side. It is a site for the formation of evaporite deposits, notably CARBONATES and sulphates. It is named after the Trucial Coast in the Persian Gulf, *sabkha* being the Arabic word for salt flat.

saccharides SUGARS (and therefore CARBOHYDRATES) divided into mono-, di-, tri- and polysaccharides. Monosaccharides are the basic units, simple sugars; disaccharides, e.g. sucrose and lactose, are formed by condensing two monosaccharides and removing water. Sucrose gives, on HYDROLYSIS, a mixture of glucose and fructose; trisaccharides comprise three basic units and polysaccharides are a large class of natural carbohydrates including STARCH and CELLULOSE.

saccharin a white, crystalline powder with about 500 times the sweetening power of sucrose. It is not very soluble in water but is used extensively as a sweetening agent, in the form of the sodium salt.

salient an interior angle in a polygon that is less than 180°.

salt a compound formed when a metal ATOM replaces one or more hydrogen atoms of an ACID (*see also* BASE).

salt dome a plug-shaped body, either circular or elongate, that is formed by the upward movement of lighter evaporitic sediments into the overlying, denser rocks. The EVAPORITE is usually HALITE. Salt domes may be 1 or 2 kilometres (1100–2190 yards) in diameter, but extend to great depths (*see also* DIAPIR).

sandstone a SEDIMENTARY ROCK comprising sand grains with sizes between 0.06 mm and 1 mm, and a variety of cements and other minerals. CALCITE is a common cement, and silica (as QUARTZ) cements sands to produce a hard sandstone often referred to as an orthoquartzite (to differentiate from the metamorphic rock, QUARTZITE). Brown and red sandstones usually have an iron-rich cement such as LIMONITE or haematite. Other minerals that may be present include FELDSPAR and MICA (*see also* ARKOSE, GREYWACKE).

saponification a process in which ESTERS are hydrolised (*see* HYDROLYSIS) by the action of acids, alkalis, boiling with water or superheated steam. If acids are used, it is the opposite process to esterification, but if alkalis are used then SOAPS result (hence the name).

saprolite weathered rock in its place of origin. The formation of saprolites depends upon a stable combination of high temperature and rainfall within a gentle landform, thus minimizing the removal of the weathered rock. Saprolites are thus best formed under tropical conditions, and large thicknesses may be generated.

saprotroph an organism that obtains its nutrition from dead and decaying organic matter. The group includes many bacteria and fungi, which are responsible for the release of nitrogen, carbon dioxide and other nutrients from the decomposing matter.

satellite any body, whether natural or manmade, that orbits a much larger body under the force of gravitation. Hence the Moon is a natural satellite of the Earth.

saturated compound a group of compounds with no double or triple BONDS; i.e. they do not form ADDITION compounds through the joining of hydrogen atoms or their equivalent.

saturated solution a SOLUTION of a substance that exists in EQUILIBRIUM with excess SOLUTE present.

Saturn (*see* APPENDIX 6 for physical data) the sixth planet in the SOLAR SYSTEM with its orbit between those of JUPITER and URANUS. The planet is mainly gaseous, with an outer zone of hydrogen and helium over a metallic hydrogen layer and a core of ice-silicate. The atmosphere is rich in methane and ethane. It is well known for its rings, which are composed of ice particles or the debris from a satellite. Saturn has 17 known satellites, Titan being the largest, and more recent discoveries include Atlas, Prometheus and Calypso (all resulting from the 1980 space probe Voyager 1).

savanna areas of grassland in tropical or subtropical zones, occupying a broad zone between the tropical forests and the semi-arid steppes. They are a result of prolonged lack of water in soils and usually occur in areas of low relief where rainfall due to mountains is absent.

scalar a quantity that has MAGNITUDE but not direction; a REAL NUMBER, as opposed to a VECTOR. The magnitude of a vector two units in length is real number or scalar 2.

scalar product the SCALAR product of two VECTORS (written A.B) is the product of the REAL NUMBERS associated with them (their MAGNITUDES) and the COSINE of the angle between their two directions, i.e. A.B = |A| |B| cos Ø.

scanning electron microscope *see* **electron microscope**.

scattering the dispersal of waves or particles upon impact with matter; applicable to light, atomic particles, etc.

scatterometer an instrument that is used to measure the light scattered from the surface of the sea. This provides information on the movement of waves and their size, and can assist in studying surface winds. The instrument is carried on a meteorological satellite.

schillerization the effect produced by the diffraction of light in the surface layers of some minerals. It appears as an iridescence.

schist a PELITIC ROCK formed by REGIONAL METAMORPHISM. Schist is the higher metamorphic grade, coarse-grained equivalent of PHYLLITE and displays a SCHISTOSITY. The minerals present vary but will usually include MICA (biotite or muscovite), QUARTZ, FELDSPAR, often GARNET, and many other accessory minerals.

schistosity the planar FABRIC (foliation) formed in a schist due to the alignment of minerals, predominantly MICAS, but also AMPHIBOLES. The alignment is generated either by the physical rotation of minerals during deformation or the metamorphic (and syntectonic—same time as the deformation) growth of new minerals that preferentially align themselves in the fabric, which is at right angles to the direction of maximum compression.

schlieren a type of FABRIC found in IGNEOUS ROCKS where aggregates of mafic minerals (iron/magnesium containing minerals that are dark in colour) become concentrated in streaks. Some may be due to XENOLITHS that have been enclosed and altered by the magma

Schottky defect named after the German physicist Walter Schottky (1886–1976), a departure from the ideal arrangement of atoms or ions in a crystal structure (*see* NON-STOICHIOMETRIC COMPOUNDS) in which an atom or ion is completely missing. Defects such as this may produce anomalous physical properties, e.g. colour, conductivity. These properties can be utilized in catalysis, semiconductors, and elsewhere.

scientific notation a useful method of writing large and small numbers. The scientific notation for a number is that number written as a power of 10 times another number, x, such that x is between 1 and 10 $(1 < x < 10)$, e.g. $145,800 = 1.458 \times 10^5$.

Schrödinger (wave) equation the basic equation used in wave mechanics, discovered by the Austrian physicist Erwin Schrödinger (1887–1961), which describes the behaviour of a particle in a force field.

scintillation small light flashes caused by ionizing radiations $(\alpha, \beta, \text{or } \gamma \text{ rays})$ striking certain phosphors, i.e. substances that luminesce. In astronomy, the twinkling of stars. The light rays from stars are effectively from point sources, and the twinkling is due to deflection by the Earth's atmosphere.

scleroproteins insoluble PROTEINS that form the skeletal components of tissues. Included are KERATIN, COLLAGEN and elastin (a fibrous protein found in lungs, artery walls, and ligaments).

scree the accumulation of mainly coarse, angular rock debris at the foot of cliffs inland. The debris is produced by the weathering and gradual disintegration of the upper slopes and cliffs, through the agencies of frost and water.

sea-floor spreading (*see also* MID-OCEANIC RIDGE) the theory, which contributed greatly to the overall concept of PLATE TECTONICS, that ocean floor is generated at mid-oceanic ridges, i.e. the margins of new tectonic plates. Basaltic magma rises along the ridge, and the newly generated crust spreads away from the ridge. On cooling, the basalts acquire a magnetism reflecting the geomagnetic field, and this gives rise to the "stripes" characteristic of MAGNETIC STRATIGRAPHY (*see also* TRANSFORM FAULT).

sea-level pressure the value of atmospheric pressure at sea level, calculated from the pressure at the measuring point and the height to sea level. The value is used on meteorological charts.

seat earth a fossil soil (palaeosol) that is found immediately beneath coal seams and represents the soil in which the vegetation grew. It is thus the last sediment deposited before plant life became established and frequently contains fossilized roots.

sea water all but 0.1 per cent of material dissolved in sea water is due to eleven components (see below). The bulk of the calcium and bicarbonate ions precipitate out as CALCIUM CARBONATE, silica is taken up by organisms, and most of the material dissolved consists of five ions: chloride, sodium, sulphate, magnesium and potassium.

ion	‰
chloride Cl⁻	19.0
sodium Na⁺	10.6

ion	‰
sulphate SO_4^{2-}	2.6
magnesium Mg^{2+}	1.3
calcium Ca^{2+}	0.4
potassium K^+	0.4
bicarbonate HCO_3^-	0.1
bromide Br^-	}
borate $H_3BO_3^-$	} less than 0.1‰
strontium Sr^{2+}	}
fluoride F^-	}

secant the function of an angle in a right-angled triangle, given by the reciprocal of the COSINE function: the secant of an angle A is 1/cosA.

second a unit of plane angle, equal to 1/60th of a minute or 1/3,600th of a degree, or π/648,000 radian. Also 1/86400th of the mean solar day.

secondary enrichment *or* **supergene enrichment** the upgrading *in situ* of ore deposits through the second stage precipitation by descending groundwater (containing dissolved sulphides and oxides), which has leached the upper layer of the ore body.

secondary metabolite a general term applied to several groups of compounds that are not directly involved with the biological proceses that contribute to growth, such as photosynthesis and respiration. However, they may be chemicals used in defence or similar mechanisms. Typical groups are TERPENOIDS, alkaloids, and the flavonoids (plant pigments).

secondary mineral a feature most common in igneous rocks (but *see also* RETROGRESSIVE METAMORPHISM), where a mineral formed at a high temperature will alter to a low-temperature mineral in the presence

351

of water. The fluid catalyses the reaction, which
most frequently produces hydrated silicate minerals, e.g. OLIVINE changing to chlorite and serpentine.
sedimentary rock one of the three main ROCK types.
Sedimentary rocks are formed from existing rocks
through processes of EROSION, denudation and subsequent deposition, compaction and cementation.
The main types are: terrigenous—derived from existing rocks on the land, e.g. sandstones and shales;
organic—produced by organic processes, e.g. limestones formed from coral reefs; chemical—precipitated from solution, e.g. EVAPORITEs such as gypsum;
and volcanogenic—associated with volcanic action,
e.g. volcanic ash deposited as TUFF, or BENTONITE.
sedimentary structure a fossil feature, preserved in
or on the bedding of a SEDIMENTARY ROCK. The structures are those generated by sedimentary processes
or the activity of organisms at the time of deposition. Structures preserved on bedding surfaces include RIPPLE MARKS, scour marks (caused by erosion),
and tool marks (caused by an object being carried
over the surface), and these are all formed by
depositional processes. SOLE MARKS are structures
preserved on the bases of beds and include trails,
tool marks, and infilled scours and load casts (a
bulging formed at the base of a bed where the upper
bed sinks into the lower while the sediment is wet,
forming a lobe). Internal sedimentary structures
include those formed by depositional processes, e.g.
lamination, convolute bedding (due to expulsion of
water from sediments deposited quickly), and slump
structures (overturned folds due to sliding sediment); organic activity, e.g. bioturbation, or by chemical activity after deposition, e.g. concretions.

seed crystal crystals that are added to a supersaturated solution (*see* SUPERSATURATION) to initiate crystallization of the whole solution.

seiche a standing wave form, produced in an enclosed body of water, which creates an apparent tide. They are usually generated by storms.

seismology the study of EARTHQUAKEs and their shock (seismic) waves to achieve a greater understanding of the deep structure of the Earth. Exploration seismology uses explosive charges to generate waves to determine structure and to search for mineral resources, e.g. hydrocarbons.

semicircular canal a structure in the inner ear in vertebrates, which helps maintain dynamic equilibrium (balance). It comprises three "loops" mutually at right angles, and in each loop is a fluid (the endolymph). When the head is moved, the fluid moves accordingly, and sensory cells respond to the movement of the endolymph. The nerve impulses are then transmitted to the brain.

semiconductor an element or compound with average resistivity (RESISTANCE related to dimensions) between that of a CONDUCTOR and an INSULATOR. The commonest are silicon, germanium, selenium, and lead sulphide. These and other materials form the basis of TRANSISTORS, DIODES, thyristors, and integrated circuits. The transistor is the basis of electronic circuits and consists of minute slices of silicon in a sandwich, altered chemically to confer differing conductivities. The passage of current through the sandwich can be controlled by a weak current through the central slice, thus creating an electrically operated switch.

senescence the process of ageing that is characterized by the progressive deterioration of tissues and the metabolic functions of their cells. According to research, senescence may be caused by the accumulation of genetic mutations within the body's cells or the expression of undesirable GENES in the later years of an individual's life. Some organisms are able to suppress senescence by REGENERATION, a process common in many simple invertebrates and one achieved by some plants using VEGETATIVE PROPAGATION.

sequence a succession of mathematical entities, x_1, x_2, ..., x_n, ..., which is indexed by the positive integers. The term x_n is the nth term or general term. A sequence may be defined by stating its nth term, e.g. a sequence whose nth term is n/n + 1 is the sequence 1/2, 2/3, 3/4, 4/5, 5/6, ..., n/n + 1, ...

series an expansion of the form $x_1 + x_2 + x_3 + ..., + x_n + ...$, where xn are real or complex numbers.

In geology, a major part of a SYSTEM, referring to the rock formed during one EPOCH (*see also* APPENDIX 5). A series can be subdivided into STAGES.

series circuit where a common current flows through the components in a circuit.

serum *see* **blood serum**.

Seyfert galaxy a small category of GALAXY that exhibits a bright nucleus (similar in many ways to a QUASAR) and weak spiral arms. They emit infrared radiation and, to a lesser extent, radio waves and X-rays (*see* ELECTROMAGNETIC WAVES).

sex chromosome one of a pair of chromosomes that play a major role in determining the sex of the bearer, with a different combination in either sex. An individual is said to be homogametic when it has a HOMOLOGOUS pair of sex CHROMOSOMES (as in the XX of female mammals) and is said to be heterogametic when it has different sex chromosomes forming its pair (as in the XY of male mammals). Sex chromosomes contain GENES that decide an individual's sex by controlling the sexual characteristics of the individual, e.g. testes in human males and ovaries, breasts, etc in human females.

sex linkage the location of a GENE on a SEX CHROMOSOME, although the expression of the gene does not necessarily affect the sexual characteristics of the individual. Some examples of sex-linked genes include red-green colour blindness and haemophilia, both RECESSIVE genes found on the X-chromosomes. Such X-linked genes cannot be passed from father to son, as the father contributes only the Y-chromosome while the mother contributes the X-chromosome to a son. The Y-chromosome contains fewer specific genes than the X-chromosome other than those responsible for maleness, and any Y-linked genes will only be inherited by male offspring.

sexual reproduction the production of progeny that have initially arisen from the fusion of male and female GAMETES in a process called FERTILIZATION. In DIPLOID organisms, sexual reproduction must be preceded by MEIOSIS to form the HAPLOID gametes if there is not to be a doubling of the number of chromosomes in all sexually reproduced offspring. As sexual reproduction involves MEIOSIS, it introduces greater genetic variation in a species, because GENETIC RECOMBINATION can occur during meiosis, with the result that any offspring will have gene combinations that differ from its parents.

shale a SEDIMENTARY ROCK that is fine-grained (composed of silt, mud and clay-sized particles). Shales are fissile due to the alignment of clay and similar minerals with their flat surfaces parallel to the planar FABRIC.

shard a volcanic glass fragment found in certain PYROCLASTIC rocks.

shear zone a zone that is usually narrow compared to its length, in which there is intense deformation due to the relative movement of two adjacent undeformed blocks. There is a continuous range of deformation from brittle (i.e. FAULTS), to ductile (as in FOLDS, etc.). Brittle faulting occurs in the upper portion of the crust (10 to 15 km/6 to 9 miles), and below this level ductile flow dominates, so that the surface expression of ductile flow at depth will be faulting. The variation in strain across a shear zone produces a progressive development in the fabric, compared to the undeformed rock. Mineral grains change shape, becoming more elongated nearer the centre of the zone and at the same time rotating to become parallel to the shear direction where deformation is most intense.

shell star a star surrounded by a layer of luminous gas. Such stars belong to SPECTRAL TYPES O, B or A.

shield a screen placed around persons or equipment to offer protection against harmful rays (eg. X-rays) or to protect an electronic component from interference from electromagnetic fields. In geology, a stable crustal area—*see* CRATON.

short-period comet a COMET with an orbit inside the SOLAR SYSTEM and a period of under 150 years, e.g. Halley's Comet.

sial a term created many years ago for the general composition of the upper crust, because it is rich in *si*lica and also *al*uminium (*see also* SIMA).

sidereal day the time taken for the Earth to make one complete rotation upon its axis with reference to the fixed stars. This is a little over four minutes shorter than the MEAN SOLAR DAY.

sidereal month the time taken by the Moon in completing one orbit of the Earth, determined using a fixed star as a reference point. It is 27 days, 7 hours, 43 minutes.

sidereal period the periods taken by the Moon and planets to reach the identical successive positions, relative to the line joining the Earth to the Sun.

sidereal time the measurement of time using the rotation of the EARTH with reference to distant stars (and not the Sun, as with "civil" time).

sidereal year the period in which the SUN appears to make one revolution with reference to the fixed stars. It is 365.26 MEAN SOLAR DAYS, slightly longer than the civil year (365.24).

sievert the SI unit of radiation dose, defined as that radiation delivered in one hour at a distance of one centimetre from a point source of one milligram of radium enclosed in platinum that is 0.5mm thick.

sigma the symbol Σ (Greek capital Sigma—"S" for "sum") used in statistics and mathematics. For example, Σx = the sum of all the values that x can assume, and Σx^2 = square each value of x then add the results.

sigma bond (σ) the bond type formed in a carbon-carbon single bond where two atomic ORBITALS overlap to form a molecular orbital surrounding the two carbon nuclei.

significant figures the digits in a number that contribute to its value, e.g. in the number 0.762 the zero is insignificant whereas the other digits are significant. If the number were 0.7620, the last zero ought to be significant because it should indicate that the number is accurate to four decimal places. However, the final zero as shown here is often added arbitrarily by the originator of the data and it may not represent such accuracy.

silica silicon dioxide (SiO_2), is one of the most important constituents of the Earth's crust. It is polymorphic (i.e. exists in more than one crystal form—when referring to elements it is ALLOTROPY) with QUARTZ (up to 573°C), tridymite (to 1470°C), and cristobalite (to 1710°C, the melting point). High-pressure forms are also known (coesite, keatite and stishovite), but these are mainly experimental products. In addition to the mineral form quartz, silica also occurs in cryptocrystalline (very finely crystalline) forms (chert, FLINT, CHALCEDONY), and as amorphous opal (the hydrated form).

Silica is used in the manufacture of glass and refractories, and the latter in the form of gannister (> 90 per cent SiO_2) is used in furnace hearths.

silica gel the hard amorphous form of hydrated silica, which is chemically inert but highly HYGROSCOPIC. It is used for absorbing water and solvent vapours, and other drying or refining tasks, and can be regenerated by heating.

silicates an enormous group of rock-forming minerals, ranging from the simple zircon, $ZrSiO_4$, which has discrete $(SiO_4)^{4-}$ anions, to highly complex structures. All are based on the SiO_4 group in a tetrahedral formation with some sharing of the oxygen atoms. Aluminium, beryllium and boron can replace silica in the $(SiO_4)^{4-}$ anion, and the CATIONS occupy the holes in the lattice. Silicate minerals are classified on the arrangement of the SiO_4 tetrahedra as follows:

neosilicates SiO_4 tetrahedra linked by cations, e.g. zircon and OLIVINE;

sorosilicates two tetrahedra sharing one oxygen e.g. epidote group;

cyclosilicates rings of 3, 4 or 6 SiO_4 sharing two oxygens e.g. BERYL;

inosilicates single chains with 2 linking oxygens e.g. PYROXENE or double chains with 2 or 3 linking oxygens e.g. AMPHIBOLES;

phyllosilicates sheets in hexagonal networks, sharing 3 oxygens e.g. MICAS;

tectosilicates frameworks in three dimensions, e.g. QUARTZ and FELDSPARS.

siliceous deposits any deposit with a high silica content, whether deposited chemically, mechanically or due to accumulation of skeletal remains (e.g. of diatoms).

silicon a non-metallic element and the second most abundant. It does not occur in native form but is found in abundance in numerous silicate minerals (*see* SILICATES) including QUARTZ itself. Silicon is manufactured by reducing SiO_2 in an electric furnace, but to get pure silicon necessitates further processing. The pure form of silicon has semi-conducting properties, and when "doped" with phosphorus or boron is used in solid-state devices.

silicon chip a SEMICONDUCTOR chip of crystalline silicon onto which is printed a microelectronic (integrated) circuit for use in computers, radios, etc.

silicones organo-silicon POLYMERS built on SiR_2O groups. Because they are inert, colourless and odourless, their uses are numerous, and the properties depend upon the degree of polymerization. They are used in oils and greases, sealing compounds, and resins (for insulation and laminates). The simpler compounds form oils while the more complex varieties form good electrical insulators.

sill an igneous INTRUSION that is concordant with the surrounding rocks and often takes advantage of weaknesses afforded by bedding planes. Sills show a range of sizes, and large ones may exceed 100 metres (330 feet) in thickness.

sima a term, complementary to SIAL, that denotes the general composition of the lower crust and upper mantle of the Earth, which is *si*licon, *i*ron and *ma*gnesium.

simple harmonic motion motion that is characteristic of many vibratory or oscillatory systems. If a point moves around a circle with a constant angular velocity, the projection of this point onto the circle's diameter (by means of a perpendicular line) produces another point, which moves back and forth along the diameter as the point moves around the circle. This is a basic illustration of simple harmonic motion. To project this moving point along an axis would produce a sine wave. Typical examples are a mass "bouncing" on a spring, a swinging pendulum, and the oscillations of air as a sound wave passes.

Simpson's rule (*also called* **parabolic rule**) a formula for approximating definite integrals, which states that the integral of a real-valued function $y = f(x)$ on an interval (a, b) is approximated by dividing the interval into an equal number of n parts at the points $x_1, x_2, ..., x_{n-1}$. The ordinates at these points are $y_1, y_2, ..., y_{n-1}$, and the width of the divisions

$h = (b - a)/2$, so the area under the curve between a and b is given by

$$1/3\ h(y_a + 4y_1 + 2y_2 + 4y_3 + ... + 2y_{n-2} + 4y_{n-1} + y_b).$$

simultaneous equations two or more equations with two or more unknown variables, which may have a unique solution, e.g. $4a - b = 10$ and $3a + 2b = 24$ yields the solution $a = 4$ and $b = 6$.

sine a function of an angle in a right-angled triangle, defined as the ratio of the length of the side opposite the angle to the length of the HYPOTENUSE.

sine rule in any triangle $A/\sin a = B/\sin b = C/\sin c$; or, any side divided by the sine of the opposite angle is equal to any other side divided by the sine of its opposite angle.

single bond *see* σ **bond**.

SI units a system of coherent metric units—Système Internationale d'Unités (*see* APPENDIX 4).

Six's thermometer a type of maximum and minimum thermometer.

skew the degree of asymmetry in a distribution curve. A skewed distribution has its modal value (the "hump" of the curve) either to the left or the right of the mean value. The degree of skewness can be measured by Pearson's COEFFICIENT of skewness, given by:

$$S = 3(\text{mean value - median value})/s$$

where s = standard deviation.

skew lines lines in space that are not parallel and do not intersect. Skew lines cannot be coplanar.

slate a low-grade fine-grained metamorphic rock. Due to the compression of deformation, sheet SILICATE minerals (e.g. MICAS, chlorite) become aligned, imparting a fissility to the rock, known as slaty CLEAVAGE. Slate is used commercially, for roofing in particular, but also larger slabs are used for other items such as billiard-table tops.

slickensides a linear feature on the plane of a FAULT, caused by the movement of one block against another. Quite often the fault plane is coated with a mineral (usually QUARTZ or CALCITE), which grows as fibrous crystals aligned to form parallel striations. The orientation of these linear features is taken to represent the direction of the most recent *net* movement, and it is thought that slickensides are created by the slow displacement of one block relative to another, at a pace that clearly permits growth of crystals.

slumping the process whereby rocks or sediments slide downslope under the inflence of gravity. It is a common occurrence in subaqueous environments with semi-consolidated sediments (*see* TURBIDITE). On land, slumping occurs on cuttings, undercut slopes, and cliffs, where there is a slip surface to facilitate movement and no obstacle in front of the slipping rock and/or soil. It occurs on spoon-shaped shear planes, which leave an arcuate trace on the ground.

snow precipitation of small ice crystals falling singly or in flakes. The crystals form from water vapour in cloud.

soap the sodium or potassium salts of the FATTY ACIDS, stearic, palmitic and oleic acid. Soaps are produced by the action of sodium or potassium hydroxide on fats (*see* SAPONIFICATION).

sodium (Na) an alkali metal that does not occur in the free state naturally because it is highly reactive. Its principal source is salt ($NaCl$) occurring in SEA WATER and salt deposits, and the metal is obtained by ELECTROLYSIS of fused NaCl. The elemental metal is silvery-white and soft, and can be cut with a knife. It reacts violently with water, and rapidly with oxygen and HALOGENS. It forms many compounds with numerous uses and is itself used as a heat-transfer fluid in reactors. Sodium is an essential element required by animals in the transmission of nerve impulses and in the maintenance of the pH of body fluids.

sodium chloride NaCl (*see* HALITE) *or* **rock salt**. A white crystalline solid obtained either from halite or the evaporation of sea water. It is used in the production of CHLORINE and in the alkali and glass industries, and is one of the most important raw materials for the chemical industries.

sodium hydroxide caustic soda, NaOH. A whitish, deliquescent (*see* DELIQUESCENCE) substance that gives a strongly alkaline solution in water. It is used a great deal in the laboratory and is a very important industrial chemical. It is used in the manufacture of soap, paper, aluminium, petrochemicals, and many other chemicals and products.

solar antapex the point on the CELESTIAL SPHERE opposite to the SOLAR APEX.

solar apex the point on the CELESTIAL SPHERE (located in the constellation Hercules) towards which the whole SOLAR SYSTEM is moving.

solar constant the sum total of the electromagnetic energy from the Sun, measured over the Earth's mean distance per unit time (currently 1.37kWm^{-2}).
In physics, it is the energy *received* on the Earth's surface, allowing for any losses due to the atmosphere.

solar corona the outer layer of the Sun's atmosphere, which reaches temperatures of 2 million K. It is a strong source of X-rays and is visible as the halo of light during an eclipse.

solar flare a temporary eruption of solar material from the Sun's surface, generating intense radio emissions and particles with very high energies. Solar flares are driven by magnetic forces.

solar parallax the average angle subtended at the Sun by the Earth's equatorial radius.

solar rotation the non-uniform rotation of the SUN, which is in the same direction as the orbital motion of the planets.

solar system the system comprising the Sun (a star of SPECTRAL TYPE G) around which are the nine planets in elliptical orbits. Nearest the Sun is MERCURY, then VENUS, EARTH, MARS, JUPITER, SATURN, URANUS, NEPTUNE, and PLUTO. In addition, there are numerous satellites, a few thousand (discovered) asteroids, and millions of comets. The age of the solar system is put at 4.5 to 4.6 billion years, a figure determined by the RADIOMETRIC DATING (uranium-lead) of IRON METEORITES. Iron meteorites are thought to be fragments of cores from early planets and thus representative of the early stages of the solar system.

solar wind the term for the stream of charged, high-energy particles (primarily ELECTRONS, PROTONS and alpha particles) emitted by the Sun. The particles travel at hundreds of kilometres per second, and the wind is greatest during flare and sunspot activity. Around the EARTH, the particles have velocities of 300–500 kms^{-1}, and some become trapped in the magnetic field to form the VAN ALLEN RADIATION BELT. However, some reach the upper atmosphere and move to the poles, producing the auroral displays (*see* AURORA).

sole mark *see* **sole structure**.

sole structure SEDIMENTARY STRUCTURES found on the base of beds formed primarily by the scouring of a current or the movement of an object (called a tool) over the sediment surface. The impression created is then often filled by sands and other sediment. Other features include flutes (tongue shapes scoured out of the mud) created by the turbulent water, and a variety of shapes caused by the passage of an object, i.e. whether it is dragged, bounced, etc. These structures help to determine the way-up of beds and can assist in derivation of the PALAEOCURRENT.

solenoid a tightly wound, cylindrical coil of wire that generates a magnetic field when current is passed through the coil.

solid in geometry, a figure with the dimensions of length, width and breadth and thus a measurable VOLUME. In chemistry, a state of matter in which the component MOLECULES, ATOMS, or IONS sustain a constant position in relation to one another, i.e. they exhibit no translational motion.

solid solution a solid solution is formed when two or more elements or compounds share a common crystalline framework. The composition may vary within finite but wide limits. Two types of solid solution are found—a substitutional solid solution, e.g. nickel/copper, where atoms of one are replaced by the other, and an interstitial solid solution, when a small atom is sited in the spaces of a lattice, between larger atoms, e.g. a carbon in metals. Several minerals exhibit solid solutions, e.g. olivine, which has iron and magnesium end members (in the formula $(FeMg)_2SiO_4$) is a substitutional solid solution.

solidus (*plural* **solidi**) in a diagram that plots temperature against composition, the solidus marks the boundary between solid and liquid/solid, at equilibrium.

solid state physics the study of all the properties of solid materials, but especially of SEMICONDUCTORS and "solid-state" devices, i.e. devices with no moving parts, as in integrated circuits, TRANSISTORS, etc.

solifluction the downhill movement of water-saturated REGOLITH. It was first described in periglacial areas (i.e. those next to a glacier or ice sheet) but now applies to all environments. It is particularly prevalent in the humid conditions of the tropics.

solstice the time at which the SUN reaches its most extreme position north or south of the equator. There are two such instants in the year.

solubility the concentration of a SATURATED SOLUTION is called the solubility of the given solute in the particular solvent used, measured in kgm^{-3}.

solubility product the number of IONS of (each type in) a compound in solution that can exist together, i.e. the product of the concentration of the ions when in equilibrium (which, for a slightly soluble salt, is a constant at a set temperature). When the solubility product is exceeded in a solution, the compound (i.e. both ions recombined) will be precipitated until the product falls to the constant value.

solute one substance dissolved in another. A solute dissolves in a SOLVENT to form a SOLUTION.

solution a single phase mixture of two or more components, which usually applies to solids in liquids and often refers to a solution in water (aqueous solution). However, other solutions include gases in liquids and liquids in liquids.

solvent a substance, usually a liquid, that can dissolve or form a SOLUTION with another substance.

somatic cell any of the cells of a multicellular organism (plant or animal) other than the reproductive cells (GAMETES).

sonar (*acronym for* Sound Navigation Ranging) a device that transmits high frequency sound and collects returning sound waves that have been reflected from submerged objects. The depth is indicated by the time taken for the return journey.

sonic boom the loud bang created by shock waves from the leading and trailing edges of an aircraft travelling supersonically. The boom results from the aircraft overtaking the pressure waves it creates ahead of itself.

sound the effect upon the ear created by air vibrations with a frequency between 20Hz and 20kHz. More generally, mechanical vibrations and waves in gases, liquids and some solids.

Southern Cross a constellation of the southern hemisphere comprising a cross of four stars, identified by its position to the west of two bright stars α and β Centauri.

Southern Lights *see* **aurora**.

space velocity a star's movement and direction in three-dimensional space.

species a group of individuals that can potentially or actually interbreed, producing viable offspring, and that within the group may show gradual morphological variations but remain different from other groups. In the taxonomic classification, species are grouped into a genus (plural *genera*), and species can themselves be subdivided into subspecies, varieties, etc. The naming of species, etc, is governed by the system of *binomial nomenclature*, so that a generic and specific name (in Latin) identify a particular individual, e.g. *panthera pardus* is the leopard.

specific heat capacity the heat required by unit mass to raise its temperature by one degree (SI units—joules per kg per Kelvin).

specific humidity (*see also* HUMIDITY) the mass of water vapour in air in proportion to the total mass of the air.

specific latent heat *see* **latent heat**.

speckle interferometry a technique for measuring small angles, e.g. the diameter of stars, which utilizes the principle of interference of light.

spectral types a classification system for stars, based upon the SPECTRUM of light they emit. The sequence is, in order of descending temperature: O—hottest blue stars; B—hot blue stars; A—blue white stars; F—white stars; G—yellow stars; K—orange stars; M—coolest red stars.

spectrochemical analysis the heating of a sample to a high temperature, producing emission lines that relate to abundance of the elements in the sample.

spectroheliograph an instrument for photographing the Sun using a single wavelength of light (i.e. monochromatic light).

spectrohelioscope a device that is essentially the same as a SPECTROHELIOGRAPH but which permits an image of the Sun to be *viewed* in light of one wavelength.

spectroscope the general term for the equipment

used in spectroscopy. It consists basically of a slit from which a beam of radiation emerges, a collimator, which renders the beam parallel, the prism that disperses the varying wavelengths, and some device for counting or measuring the radiation.

spectrum (*plural* **spectra**) the range of WAVELENGTHS obtained upon resolution of ELECTROMAGNETIC radiations. An obvious example is the coloured "rainbow" bands obtained when white light passes through a prism.

speed for a body moving in a straight line or continuous curve, the ratio of distance covered to the time required to cover that distance. Units vary, e.g. metres per second (ms^{-1}), miles per hour (mph).

speed of light as revealed in the theory of RELATIVITY, a universal and absolute (independent of the speed of the observer) value that is 2.998×10^8 ms^{-1}, or 186,281 miles per second.

speed of sound the value for the speed (VELOCITY) of sound depends upon the nature of the medium and the temperature. In air at 0°C, the speed is 332 ms^{-1}, or about 760 mph. In fresh water, the speed is 1410 ms^{-1}.

sphere a circular solid figure with all points on its surface an equal distance from the centre. In two-dimensional CARTESIAN CO-ORDINATES, the equation of a sphere is $(x - a)^2 + (y - b)^2 + (z - c)^2 = r^2$. For a sphere of radius r, vol = $4/3\pi r^3$; surface area A = $4\pi r^2$.

spherical co-ordinates (*also called* **spherical polar co-ordinates**) a three-dimensional polar co-ordinate system in which the position of a point P is given by the co-ordinates (r, θ, φ), where r is the radius vector with respect to the origin O of two axes at right angles; θ is the angle between the vertical (polar) axis and the radius vector; and φ is the angle between the horizontal axis (x-axis) and the projection of the radius vector on the horizontal plane.

spilite a low-grade METAMORPHIC ROCK formed by METASOMATISM of MID-OCEAN RIDGE BASALTS (through the agency of sea water). The sea water becomes warm due to the intrusive activity and reacts with the rocks, introducing sodium and altering the basaltic mineral assemblage to that of spilite, *viz.* chlorite, plagioclase FELDSPAR, actinolite (an AMPHIBOLE), CALCITE, sphene (a calcium titanium silicate), with accessory minerals.

spiral galaxy a galaxy that has spiral arms containing the stars, dust and gas. The SOLAR SYSTEM belongs to such a galaxy.

spontaneous combustion the ignition of a substance without application of a flame. It may occur through the production of heat from slow OXIDATION within the substance.

spontaneous generation a now discredited theory that living organisms could arise from non-life. It was believed that spontaneous generation could occur in, for example, rotting meat or fermenting broth, giving rise to an individual organism, but it is now known that all new organisms originate from the parent organism from whom they have inherited a genetic ancestry.

spore a small reproductive unit, usually consisting of one cell, that detaches from the parent and disperses to give rise to a new individual under favourable environmental conditions. Spores are particularly common in fungi and bacteria but also occur in all groups of green land plants such as ferns, horsetails and mosses.

square *see* **polygon**.

stage a subdivision of the SERIES, and a unit in chronostratigraphy that is equivalent to an age in geologic time units. It is the rock accumulated during one age.

stalactite a hanging deposit of calcium carbonate ($CaCO_3$) formed from the roof of a cave by drips of calcium-rich solutions. Stalactites resemble icicles.

stalagmite an upstanding growth of calcium carbonate ($CaCO_3$) formed on the floor of a cave by drips of calcium-rich solutions. Often found with stalactites, when the two forms may eventually meet and join.

standard time the time in the time zones established by international agreement. Each zone is equal to one hour and is approximately 15° of longitude wide.

standing wave a disturbance produced when two similar wave motions are transmitted in opposite directions at the same time. This results in INTERFERENCE, with the combined wave effects producing maxima and minima over the area of interference. The resultant waveform is contained within fixed points and does not move, hence standing or stationary wave.

star a body of matter, similar to the SUN, which is contained by its own gravitational field. Stars are glowing masses that produce energy by thermonuclear reactions (NUCLEAR FUSION). The core acts as a natural nuclear reactor, where HYDROGEN is consumed and forms HELIUM with the production of electromagnetic radiation (*see* ELECTROMAGNETIC WAVES).

starch a polysaccharide (*see* SACCHARIDES) found in

all green plants. It is built up of chains of GLUCOSE units arranged in two ways, as amylose (long unbranched chains) and amylopectin (long cross-linked chains). Potato and some cereal starches contain about 20-30 per cent amylose and 70-80 per cent amylopectin.

static electricity stationary electric charges which result from the electrostatic field produced by the charge (*see also* VAN DE GRAAFF GENERATOR).

statics a branch of mechanics that studies the combination of forces in equilibrium, i.e. the behaviour of matter under applied forces when there is no resulting motion.

stationary orbit the orbit of a satellite around a body such that it holds a fixed point about the latter's equator. For the Earth, the GEOSTATIONARY orbit is approximately 36,000 kilometres (22,370 miles) above the equator. This is a property employed in the positioning of communications satellites.

statistics the branch of mathematical science dealing with the collection, analysis and presentation of quantitative data, and drawing inferences from data samples by the use of probability theory.

stellar evolution the various stages of the life of a star, which begins with the creation of the star from the condensation of gas, primarily hydrogen. The growth of the clouds pulls in more gas, and the increase in gravity compresses the molecules together, which attracts more material and creates a denser mass. The heat normally produced by molecules due to their vibratory motion is increased greatly, and the temperature is raised to millions of degrees, which facilitates NUCLEAR FUSION. The supply of hydrogen continues to be consumed (and the star occupies the MAIN SEQUENCE of the HERTZSPRUNG-RUSSELL DIAGRAM) until about 10 per cent has gone, and then the rate of combustion increases. This is accompanied by collapses in the core and an expansion of the hydrogen-burning surface layers, forming a RED GIANT. Progressive gravitational collapses and burning of the helium (generated by the consumption of the hydrogen) result in a WHITE DWARF, which is a sphere of enormously dense gas. The white dwarf cools over many millions of years and forms a black dwarf—an invisible ball of gases in space. Other sequences of events may occur, depending upon the size of the star formed. BLACK HOLES and NEUTRON STARS may form from red (super) giants via a SUPERNOVA stage.

stellar wind a similar phenomenon to the SOLAR WIND—an outpouring of material from a hot star.

stereochemistry the part of chemistry that covers the spatial arrangement of atoms within a molecule (*see* ISOMER).

stereoisomerism isomerism due to the different arrangement in space of atoms within a molecule, giving ISOMERs that are mirror images of each other.

stereotaxis the movement or reaction of an organism in response to contact with a solid body.

steric hindrance the phenomenon whereby the arrangement in space of atoms in reacting molecules hinders or slows a chemical reaction.

steroids a group of LIPIDS with a characteristic structure comprising four carbon rings fused together. The group includes the sterols (e.g. CHOLESTEROL), the BILE acids, some HORMONES, and vitamin D. Synthetic steroids act like steroid hormones and include derivatives of the glucocorticoids used as anti-inflammatory agents in the treatment of rheumatoid arthritis; oral contraceptives which are commonly mixtures of OESTROGEN and a derivative of PROGESTERONE (both female sex hormones); anabolic steroids e.g. TESTOSTERONE, the male sex hormone, which is used to treat medical conditions such as osteoporosis and wasting. However, much publicity surrounds the use of the anabolic steroids by athletes, contrary to the rules of sports-governing bodies, to increase muscle bulk and body weight.

stoichiometry an aspect of chemistry that deals with the proportion of elements (or chemical equivalents) making pure compounds.

stony meteorites meteorites composed mainly of rock-forming SILICATES, including PYROXENE, OLIVINE and plagioclase (see FELDSPAR) with some nickel-iron. This type accounts for the vast majority of meteorites that are *seen* to fall. They are termed either chondrites or achondrites, depending upon the presence, or lack, of chondrules, which are glassy droplets up to 2 mm in size, produced by the melting and sudden cooling of silicate material.

strain when forces acting upon a material produce distortion, it is said to be strained, or in a state of strain. Strain is represented as a ratio of the change in dimension or volume to the original dimension or volume.

strain gauge a device to measure strain. A basic version comprises a fine metal or semiconductor grid on a backing sheet, which is fastened to a body that is to be subjected to strain. The grid is then strained by the same amount, altering the electrical properties of the grid, which can be measured. There are many other strain gauges, which employ light, vibrations/sound and the use of liquid crystals.

stratigraphic break a break in the geological record denoting no deposition or the dominance of erosion and represented by unconformities and gaps in the sequence.

stratigraphic unit a well-defined and separate rock unit, which can be characterized in several ways: a lithological distinction—lithostratigraphic; on fossil content—biostratigraphic; or by time span—chronostratigraphic.

stratigraphy the study of (stratified) rocks, with particular regard to their position in time and space, the designation of type sections (i.e. a rock succession at a locality, which is used as the standard to which other successions are matched), and the correlation of like successions. This may entail the use of fossils, characteristic lithologies, and/or time intervals.

stratocumulus a grey or white CLOUD composed of sheets or layers usually with dark patches.

stratopause the top of the STRATOSPHERE at about 50 kilometres.

stratosphere the layer of the atmosphere above the TROPOSPHERE, which stretches from 10 to 50 kilometres (6 to 31 miles) above the ground. It is a stable layer with the TROPOPAUSE at the base. The temperature increases from the lower part to the upper, where it is 0°C, and the higher temperatures are due to ozone-absorbing ultraviolet radiation. The inversion of temperatures creates the stability that tends to limit the vertical extent of cloud, producing the lateral extension of, for example, CUMULONIMBUS cloud into the anvil head shape.

stratum (*plural* **strata**) a layer or bed of rock that has no limit on its thickness.

stratus a spread-out cloud form, with an even base and generally grey appearance, through which the Sun may be seen, providing the cloud is not too dense. It occasionally forms ragged patches.

stress force per unit area. When applied to a material, a corresponding STRAIN is created. The two main types are tensile (or compressive), and shear stress. Units, typically, include: kNm^{-2}, $lbfin^{-2}$.

striation *or* **striae** a product of glaciation that results in lines or grooves being scratched in exposed rock due to the action of hard rocks embedded in the base of the glacier. The lines may provide a guide to the direction of ice movement in areas that have undergone glaciation.

strike the direction of a horizontal line, measured with a compass, on an inclined plane, e.g. a rock bedding plane. It is perpendicular to the DIP. Measurement of strike and dip is a vital procedure for geologists undertaking mapping, as it allows them to plot the results and determine large-scale trends and structures.

strike-slip fault a FAULT in which the displacement is parallel to the STRIKE of the fault plane. The movement on such a fault can essentially be in one of two directions, called dextral and sinistral. The latter, sinistral, is when one side is moved to the left of the other, from where the observation is made.

stroboscope an instrument used to view rapidly moving objects by shining a flashing light source, of variable periodicity, at the object. Synchronization of the two frequencies renders the object apparently immobile.

stroma (*plural* **stromata**) any tissue that functions as a framework in plant cells (*see* CALVIN CYCLE).

stromatolite laminated calcareous structures produced by calcareous algae (cyanophyte or cyanobacteria), which trap and bind fine sediment to produce, in most cases, a structure with some vertical dimension, e.g. domes, columns. Stromatolites are found today in tropical carbonate environments in shallow water, but in the past their distribution was more widespread. Current forms have laminae millimetres thick and an overall length of centimetres, while Precambrian varieties are known that are several metres thick.

strophism the twisting, by progressive growth, of a stalk in response to a stimulus from a particular direction, e.g. light.

structural formula a formula providing information on the ATOMS present in a MOLECULE and the way that they are bound together, i.e. an indication of the structure.

strychnine a crystalline ALKALOID with a very strong bitter taste and a very dangerous action upon the nervous system.

subduction zone an essential component of the concept of PLATE TECTONICS. It is the area where a lithospheric plate descends beneath continental crust at deep ocean trenches, thus returning old LITHOSPHERE material to the MANTLE. It is therefore known as a destructive plate boundary. Evidence for the process is afforded by MAGNETIC STRATIGRAPHY of the ocean floor. At a constructive boundary (MID-OCEANIC RIDGE) where spreading occurs, parallel and similar magnetic stripes occur on either side of, and concordant with, the ridge. At a subduction zone, the stripes are discordant, and stripes of various ages occur along the boundary because the remainder have been subducted. A zone of earthquake foci is associated with the subducted slab. Volcanism is also a feature found near subduction zones, in ISLAND ARCS.

sublimation the production of a vapour directly from a solid, without going through the liquid phase.

substellar point the point on the Earth where a particular star would be vertically overhead, and defined by a line from the centre of Earth to the star. Where the line cuts the surface is the substellar point.

substitution (reaction) a reaction in which an atom or group in a molecule is replaced by another atom or group, often hydrogen by a HALOGEN, hydroxyl, etc.

substrate in biology, (1) the surface upon which an organism lives and from which it may derive its food. (2) a substance/reactant in a reaction which is catalysed by an ENZYME.

In electronics, the single crystal of SEMICONDUCTOR used as the base upon which an integrated circuit or TRANSISTOR is printed.

sucrose a disaccharide CARBOHYDRATE ($C_{12}H_{22}O_{11}$) occurring in beet, sugar cane and other plants (*see* SACCHARIDES).

sugar a crystalline monosaccharide or oligosaccharide (a small number, usually two to ten, monosaccharides linked together, with the loss of water), soluble in water. The common name for SUCROSE.

sulphur a yellow non-metallic element that exhibits ALLOTROPY. It is widely distributed both in the free state and in compounds, especially as sulphates (e.g. GYPSUM) and sulphides (e.g. PYRITE), and as hydrogen sulphide (H_2S) in natural gas and oil. It is manufactured by heating pyrite or purifying the naturally occurring material. Its primary use is in the manufacture of SULPHURIC ACID, but it is also used in the preparation of matches, fireworks, dyes, fertilizers, fungicides, etc.

sulphuric acid (H_2SO_4) a strong acid that is highly corrosive and reacts violently with water, with the generation of heat. It is manufactured by the CONTACT PROCESS. It is used widely in industry in the manufacture of dyestuffs, explosives, other acids, fertilizers, and many other products.

summation convention an abbreviated notation, used particularly in tensor analysis and RELATIVITY THEORY, in which a product of tensors is to be summed over all possible values of any index that appears twice in the expression.

Sun the star nearest to Earth and around which Earth and the other planets rotate in elliptical orbits. The Sun has a diameter of 1.392 x 10^6 kilometres, and its mass is approximately 2 x 10^{30} kilograms. The interior reaches a temperature of 13 million degrees Centigrade, while the visible surface is about 6000°C. The internal temperature is such that thermonuclear reactions occur, converting HYDROGEN to HELIUM with the release of vast quantities of energy. The Sun is approximately 90 per cent hydrogen, 8 per cent helium, and is 5 million years old—roughly halfway through its anticipated life cycle.

sunshine recorder *or* **Campbell-Stokes recorder** an apparatus comprising a glass sphere that focuses the Sun on card marked with hours. The focusing

creates sufficient heat to burn a track on the card, thus
recording the duration of the sunshine.

sunspots the appearance of dark areas on the surface of the SUN. The occurrence reaches a maximum
about every eleven years. Sunspots have intense
magnetic fields and are associated with magnetic
storms and effects such as the *aurora borealis* (*see*
AURORA). The black appearance is due to the sunspots dropping in temperature, to about 4000K.

superfluidity is when a fluid flows without friction,
e.g. helium-4 (^{4}He) below 2.19K behaves in this way,
with zero effective viscosity. The temperature is
called the lambda point, and at temperatures above
^{4}He is termed helium I and helium II below.

supergene enrichment *see* **secondary enrichment**.

supernova (*plural* **supernovae**) a star that ex-
plodes, it is thought, because of the exhaustion of its
hydrogen (*see* SUN), whereupon it collapses, generating high temperatures and triggering thermonuclear reactions. A large part of its matter is flung
into space, leaving a residue that is termed a WHITE
DWARF star. Such events are very rare, but at the
time of explosion, the stars become one hundred
million times brighter than the sun.

superconductivity the property of some metals and
alloys whereby their electrical RESISTANCE becomes
very small around ABSOLUTE ZERO. The potential uses
for **superconductors** include circuits in large computers, where superconductive circuits would generate far less heat than is the case presently. This
would enable larger and faster computers to be
built. Transmission of electricity and reduction of
the associated heat loss is also under study.

superposable denoting a property of a geometrical

figure that allows its co-ordinates to be transposed to coincide with those of another figure.

superposition, law of a law, first proposed in the 17th century, that rocks are deposited in successive layers. Thus, in an *undisturbed* succession of SEDI-MENTARY ROCKS, each layer overlies an older one, i.e. the layers become younger going up the succession.

supersaturation the state of a SOLUTION when it contains more dissolved SOLUTE than is required to produce a SATURATED SOLUTION.

surd the root of quantity that can never be exactly expressed because it is an IRRATIONAL NUMBER, e.g. $\sqrt{3}$ = 1.73205...

surface integral the integral of a function of several variables with respect to the surface area over a surface in the domain of a function.

surface tension a tension created by forces of at-traction between molecules in a liquid, resulting in an apparent elastic membrane over the surface of the liquid.

surface wind the wind as measured close to the Earth's surface, actually at a height of 10 metres (32.8 ft). The speed of the surface wind is reduced by friction of the surface.

surfactant (*also called* **surface-active agent**) a compound that reduces the SURFACE TENSION of its SOLVENT, e.g. a detergent in water.

suspension a two-phase system with particles dis-tributed in a less dense liquid or gas. Settling is prevented or slowed by the VISCOSITY of the fluid, or impacts between the molecules of the fluid and the particles themselves.

syenite a coarse-grained IGNEOUS ROCK with alkali FELDSPAR and ferromagnesian minerals, including

BIOTITE, hornblende (AMPHIBOLE) and aegirine-augite (PYROXENE). The feldspar constitutes approximately two-thirds of the rock. Syenites occur as INTRUSIONS on continental crust and in ring complexes, i.e. a complex mix of intrusions including RING DYKES, PLUGS and cone sheets (a dyke with a conic cross-section).

symbiosis a relationship between organisms, usually two different species, which has beneficial consequences for at least one of the organisms. There are various forms of symbiosis, including commensalism, where one party benefits but the other remains unharmed, and parasitism, where one party greatly benefits (the PARASITE) at the other party's expense. Symbiosis can also solely refer to mutualism, where both parties benefit and neither is harmed.

symbol a letter or letters that represent an element or an atom of an element (*see* element table in APPENDIX 2 for list of symbols).

symmetry the property of a geometrical figure whose points have corresponding points reflected in a given line (axis of symmetry), point (centre of symmetry, e.g. a circle) or plane (reflection).

synapse (*see also* NEURON) the junction between two nerve cells where a minute gap (of the order of 15 NANOMETRES) occurs. The nerve impulse is carried across the gap by a neurotransmitter substance (*see* ACETYLCHOLINE). The chemical diffuses across the gap connecting the axon of one nerve cell to the dendrites of the next. An individual neuron commonly has several thousand such junctions with around 1000 other neurons.

syncline a FOLD shaped like a basin, with the younger strata uppermost (in the centre).

synergism is when the combined effect of two substances (e.g. drugs) is greater than expected from their individual actions added together.

synthesis the formation of a compound from its constituent elements or simple compounds.

system a rock succession formed during a period, i.e. the chronostratigraphic equivalent of the period time unit.

systems analysis the application of mathematics to the analysis of a particular system, involving the construction of a mathematical model, which is then analysed and the results applied to the original system.

systolic blood pressure the pressure generated by the left VENTRICLE of the HEART at the peak of its contraction. Since the left ventricle has to pump blood to all parts of the body, it generates a higher pressure than the right ventricle, which pumps blood only to the lungs. In normal people, the systolic blood pressure is 120 mm of mercury (120 mm Hg), and when the ventricle relaxes, pressure is still maintained in the blood vessels. This resting pressure is called the diastolic pressure and is approximately 80 mm Hg. This is the familiar blood pressure measurement and is represented as 120/80. This fluctuation in pressure is responsible for the pulse, which also represents the heartbeat.

T

talc an hydrated magnesium silicate ($Mg_3Si_4O_{10}(OH)_2$), which is commonly secondary in origin. It is formed from the alteration of OLIVINE, PYROXENE and AMPHIBOLE, and also from the metamorphism of siliceous DOLOMITES. It is used as a mineral filler, and in purified form in toilet preparations.

taluvium a deposit on a sloping hill consisting of coarse and fine material. The term originates from *talus* (a slope of SCREE) and *colluvium* (finer weathered rock debris that has moved downslope by creep or through the agency of surface water).

tangent a function of an angle in a right-angled triangle, defined as the ratio of the side opposite the angle to the length of the side adjacent to it. In geometry, a straight line that just touches the circumference of a circle.

tangent rule tan 1/2 (A – B)/tan 1/2 (A + B) = a – b/a + b.

tannic acid a polymeric ESTER-like derivative of gallic acid and GLUCOSE. It is used as a MORDANT in dyeing and in tanning and ink manufacture.

tartaric acid an organic acid which exists in four forms that are STEREOISOMERS. The commonest form (d-tartaric acid) is used in dyeing, the manufacture of baking powder and "health salts." Another form (dl-tartaric acid, or racemic acid) occurs in grapes.

384

tautomerism a special case of structural isomerism (*see* ISOMER), often called dynamic isomerism. It is when a compound exists in a mixture in two forms, or isomers (in equilibrium). The two forms can each change to the other as when, for example, one isomer is extracted from the mixture, then some of the other isomer will change to re-establish equilibrium. The ability to make this reversible change is conferred by a mobile atom or group, often hydrogen, which alters its position in the molecule possibly with the changing of a double band. Isomers produced in this way are called **tautomers**, and each can give rise to stable derivatives. A typical example would be:

$$> \overset{\overset{\textstyle H}{\textstyle |}}{C} - C = C < \; = \; > C = C - \overset{\overset{\textstyle H}{\textstyle |}}{C} <$$

taxis (*plural* **taxes**) the movement of a cell or organism in response to a stimulus in environment. This stimulus may be temperature (thermotaxis), light (phototaxis), gravity (geotaxis), or chemical (CHEMOTAXIS).

taxonomy the study, identification and organization of organisms into a hierarchy of diversity according to their similarities and differences. Taxonomy is concerned with the classification of all organisms, whether plant or animal, dead or alive, e.g. fossils are also important. Modern taxonomy provides a convenient method of identification and classification of organisms, which expresses the evolutionary relationships to one another.

T-cell a type of white blood cell (LYMPHOCYTE) that differentiates in a gland called the thymus, situated in the thorax. There are a whole variety of T-cells involved in the recognition of a specific foreign body (ANTIGEN), and they are particularly important in combating viral infections and destroying bacteria that have penetrated the cells of the body.

tectonic concerned with earth movements, as involved in FOLDing and FAULTing.

tectosilicates one of the rock-forming silicates and itself an important group that contains QUARTZ and the FELDSPARS (also the FELDSPATHOIDS and ZEOLITES). Tectosilicates consist of numerous SiO_4 tetrahedra linked at each of their four corners so every oxygen ion is shared between two tetrahedra, thus creating a framework. This produces a silicon to oxygen ratio of 1 : 2.

telophase the last stage of MEIOSIS or MITOSIS in EUCARYOTIC cells. During telophase, a nuclear membrane forms round each of the two sets of CHROMOSOMES that have formed separate groups at the spindle poles. The chromosomes decondense, the nucleoli reappear, and the cell eventually splits to form two daughter cells.

temperature degree of heat or cold against a standard scale.

temperature inversion *see* **inversion**.

tensor in mathematics, a generalized form of VECTOR. In biology, a muscle that stretches a part of the body without altering the relative position of that part.

tephra a general term that is applied to all particles or fragments blown out during the explosive eruption of a VOLCANO. It applies to all solid matter, irrespective of size or shape, although it usually applies to air-fall material and not PYROCLASTIC flows.

Tephra deposits may remain unlithified (i.e. are not consolidated to form rocks) for long periods and are subject to rapid erosion.

terminal velocity the constant VELOCITY achieved by a body falling through a medium when the pull of GRAVITY is equalled by the frictional resistance.

terpenes colourless, liquid HYDROCARBONS occurring in many fragrant natural oils of plants. The general formula is $(C_5H_8)n$ where C_5H_8 is the basic isoprene unit. This leads to their classification: monoterpenes are $C_{10}H_{16}$; sesquiterpenes $C_{15}H_{24}$; diterpenes $C_{20}H_{32}$, and so on.

terpenoids a group of plant SECONDARY METABOLITES based on isoprene units (C_5; *see also* TERPENES). The group includes many essential oils, the CAROTENOIDS, rubber, and the gibberellins, which are plant growth substances promoting shoot elongation in some plants and promoting seed germination.

terrigenous sediments sediments produced by erosion of, or deposition on, the land; sediments often collectively termed siliciclastic, i.e. composed of silicate minerals and rock fragments.

testosterone a male sex hormone that promotes the development of male characteristics.

thermal conductivity a measure of the rate of heat flow along a body by conduction.

thermal metamorphism *or* **contact metamorphism** the process whereby rocks intruded by an igneous body are recrystallized in response to the heat. There is little or no associated pressure, thus large-scale structures may remain while the mineralogy changes. METASOMATISM often accompanies thermal metamorphism. The zone of altered rock is called the metamorphic AUREOLE, and clearly the greatest changes are observed next to the intrusion. PELITIC rock assemblages show the greatest sensitivity to the temperature, and new minerals will form, e.g. cordierite (a silicate with aluminium, iron and magnesium), and andalusite (aluminium silicate, Al_2SiO_5).

thermistor a temperature-sensitive SEMICONDUCTOR whose RESISTANCE decreases with temperature increase. Thermistors are used for temperature measurement and compensation.

thermochemistry the branch of chemistry dealing with the heat changes of chemical reactions.

thermocouple a device for measuring temperature, comprising two metallic wires joined at each end. The temperature is measured at one join, and the other join is kept at a fixed temperature. A temperature difference between the two joins creates a thermoelectric e.m.f. (electromotive force = VOLTAGE), which causes a current to flow. Either the voltage or the current can be measured, thus creating a calibrated device.

thermodynamics the study of laws affecting processes that involve heat changes and energy transfer. There are essentially three laws of thermodynamics. The first (the law of conservation of energy) states that within a system, energy can neither be created nor destroyed. The second law says that the ENTROPY of a closed system increases with time. The result of the third law (or NERNST heat theorem) is that absolute zero can never be attained.

thermograph a thermometer that records continuously, and where the recording is usually achieved by means of an electrical device.

thermography the medical scanning technique

whereby the INFRARED RADIATION or radiant heat emitted by the skin is photographed, using special film, to create images. An increase in heat emission signifies an increase in blood supply, which may be indicative of a CANCER. The technique is used to detect cancers, especially of the breast.

thermoluminescence a phenomenon whereby a material emits light upon heating due to ELECTRONS being freed from DEFECTs in crystals. The defects are generally due to ionizing radiations, and the principle is applied in dating archaeological remains, especially ceramics, on the assumption that the number of trapped electrons, caused by exposure to radiations, is a function of time. Although this is not absolutely correct, an estimate of the age of a piece of pottery can be obtained by heating and comparing the thermoluminescence with that of an item of known age.

thermolysis breakdown of a compound or molecule by heat.

thermometer an instrument used to measure temperature. Any property of a substance, providing it varies reliably with temperature, may form the basis of a thermometer. This includes expansion of liquids or gases, or changes in electrical RESISTANCE.

thermoplastic a plastic material that can be melted or softened by heat and then cooled, repeatedly, without significant alteration in its properties, e.g. polystyrenes, vinyl polymers.

thermostat a device for maintaining a constant temperature through the supply or non-supply of heat when the required temperature is not achieved.

thermotropism the growth of a plant in the direction of a heat source.

thin layer chromatography CHROMATOGRAPHY that occurs essentially in two dimensions, with the separation of small quantities of mixture achieved by movement of a solvent across a flat surface on sheets of absorbent paper or special materials (e.g. silica gel supported on glass plates). The components of the sample move at different rates across the "plate" because of differences in solubility, size, charge, etc. After the process, the components can be examined *in situ* or removed for further analysis. The technique has its disadvantages, but it is used widely for checks on purity or to characterize complex materials.

thin section a thin slice of mineral or rock, on a glass slide, used for study under the microscope. The chip of material is ground in several stages to reach a thickness of 0.03 mm, thus permitting light to be transmitted through it when it is examined with the polarizing microscope. It is an important technique in PETROGRAPHY.

thixotropy the property of some fluids to be very viscous until a STRESS is applied, when the fluid flows more easily. This principle is utilized in nondrip paints.

Thornthwaite climate classification (*see also* KÖPPEN CLASSIFICATION) a classification of climates devised in the early 1930s by C.W. Thornthwaite and based upon characteristic vegetations and the associated precipitation. Two parameters were developed: P/E and T/E. P/E represents the total monthly precipitation divided by the total monthly evaporation, and T/E is similar save that it refers to the temperature. Based upon the precipitation figures, five humidity provinces were created:

Province	Vegetation	P/E
A	rain forest	>127
B	forest	64–127
C	grassland	32–63
D	steppe	16–31
E	desert	< 16

Each province can be further subdivided by the rainfall pattern, e.g. abundant in all seasons, deficient in winter. The T/E figures provide an additional classification into provinces:

	Province	T/E
A'	tropical	128
B'	mesothermal	64–127
C'	microthermal	32–63
D'	taiga	16–31
E'	tundra	1–15
F'	frost	0

With all this information, Thornthwaite produced a climatic world map of 32 climate types. Each of these areas can be associated with certain soils, geomorphological processes, etc, related to and depending upon the climate.

three-body problem for three bodies that show mutual attraction there is no general solution to compute their behaviour, although specific solutions are known.

throw the vertical component of the displacement in a dip-slip FAULT, i.e. in faults where the displacement is parallel to the fault plane.

thrust a low-angle FAULT (usually less than 45°) in which the sense of movement is reverse.

thunder the rumbling noise that accompanies lightning flashes and that is due to violent thermal

391

changes caused by the electrical discharge. The continuing noise is due to sound travelling from the various sections of the discharge—because the spark can be several kilometres long.

thymine ($C_5H_6N_2O_2$) a nitrogenous base component of DNA that has the structure of a PYRIMIDINE. Thymine always base-pairs with ADENINE in a DNA molecule, but in RNA molecules it is replaced with URACIL.

thyroid gland *see* **endocrine system**.

tide tides affect the surface layers of a planet (or natural satellite) whether liquid or solid, due to the effects of gravitational forces. The ocean tide on Earth is due to the attraction of mainly the Moon, but also the Sun. Variation in tides is caused by the positions of the three bodies and when the Moon and the Sun "pull" in the same direction, there is a high spring tide; when they are at 90°, there is a low neap tide. Other factors that affect tides are the uneven distribution of water on the Earth's surface, and the topography of the sea bed.

till sediment deposited due to the action of glacial ice without water as an agent. The size varies from clay particles to rock fragments. There is a variety of tills depending upon their method of release. For example, subglacial melting gives lodgement till, and the thawing of stationary ice produces melt-out till.

tillite a consolidated deposit of till (or BOULDER CLAY).

tin a soft, malleable and ductile metal (SYMBOL Sn) which exhibits ALLOTROPY. It occurs naturally as oxides and is used to coat steel and in producing alloys (solders, fusible alloys, etc.).

tissue a group of cells with a similar function, which aggregate to form an organ.

tissue culture the culture or growth, outside the body, of TISSUES of living organisms. The cells are placed in an artificial medium containing nutrients, and other factors such as temperature and pH are controlled, while waste products are removed. The technique has proven valuable in studying cell growth, and it has been applied to the propagation of plants.

titanium (Ti) a malleable and ductile metal that resembles iron. The main source is the ore rutile (TiO_2). It is characterized by lightness, strength and high resistance to corrosion. It is therefore useful in aircraft and missile manufacture.

titration the laboratory procedure of adding measured amounts of a SOLUTION to a known volume of a second solution until the chemical reaction between them is complete, enabling the unknown strength of one solution to be determined.

tomography a scanning technique that uses X-rays for photographing particular "slices" of the body. A special scanning machine rotates around the horizontal patient, taking measurements every few degrees over 180°. The scanner's own computer builds up a three-dimensional image that can then be used for diagnosis. Such a technique has the dual benefit of providing more detail than a conventional X-ray and yet delivers only one fifth of the dose.

tonne a metric ton (1000kg).

torus a "doughnut" shaped ring, generated by rotating a circle about an axis.

totipotency the capacity of a cell to generate all the characteristics of the adult organism. Totipotent cells have full genetic potential, unlike most adult cells, which have lost this during the process of differentiation when they form cells with specialized functions.

trace element an ELEMENT that occurs in very small quantities in rocks, but which can be detected by geochemical analysis. All elements but the most commonly occurring form trace elements and in quantities much less than 1 per cent. Very often trace elements occur in minute quantities ranging from a few parts per million to several hundred (*see also* X-RAY FLUORESCENCE SPECTROMETRY).

trace fossil a SEDIMENTARY STRUCTURE due in some way to the presence or activity of an organism. The study of trace fossils is called ichnology. They are most common at the junction of different lithologies (*see* LITHOLOGY), for example shale and sandstone. The fossils may occur as ridges, tubes, burrows, etc.

trachyte a fine-grained volcanic IGNEOUS ROCK. It is an intermediate alkaline rock high in alkali metals (e.g. lithium, sodium, potassium) and alkaline earth metals (e.g. magnesium, calcium, strontium). Ferromagnesian minerals will also be present, the types being dictated by the rock composition. SYENITE is the coarse-grained equivalent of trachyte.

trade winds steadily blowing easterly winds that blow from subtropical high pressure areas between latitudes 30° north and south. The winds are generally northeasterly in the northern hemisphere and southeasterly in the southern hemisphere. Associated weather patterns include fine weather to the north and east, and stormy conditions to the west and near the equator.

transcription the formation of an RNA molecule from one strand of a DNA molecule. Transcription involves many processes, starting with the unwinding of the double-stranded DNA helix, along which an enzyme, called RNA polymerase, travels and catalyses the formation of the RNA molecule by pairing NUCLEOTIDES with the corresponding sequence of the DNA strand. As the RNA molecule leaves, the DNA reforms its double-stranded helix.

transducer a device that converts one form of energy into another, often a physical quantity into an electrical signal, as in microphones and photocells. The reverse also applies, as in loudspeakers.

transferase an ENZYME that catalyses the transfer of chemical groups between molecules, e.g. acyl transferase.

transfer RNA (tRNA) one of the three major classes of RNA that functions as the carrier of AMINO ACIDS to RIBOSOMES, where the POLYPEPTIDE chains of PROTEINS are formed. Every tRNA molecule has a structure that will accept only the specific attachment of one amino acid.

transformation a rearrangement of the term of a mathematical expression. The changing (or mapping) of one shape into another by reflection, rotation, dilation, translation, etc.

transformer a device for changing the VOLTAGE of an ALTERNATING CURRENT. The unit consists essentially of an iron core with two coils of wire. Current fed into the primary coil generates a current in the secondary through ELECTROMAGNETIC INDUCTION. The ratio of the voltage between the coils is determined by the ratio of the number of turns in each coil.

transform fault a fundamental component of the theory of PLATE TECTONICS because transform faults permit the subdivision of the Earth into plates undergoing relatively little internal deformation.

Viewing the MID-OCEANIC RIDGES, numerous transform faults appear to offset the ridge repeatedly.
However, the lateral displacement on one side of the
fault is actually taken up by the formation of new
crust at the ridge (or by shortening at a trench). In
effect, the sense of movement on the fault is reversed. Thus, a fracture that was once regarded as
displaying left lateral movement (i.e. when the
block to the left of the fault is moved with respect to,
and when viewed from, the block on the right of the
fault) was reinterpreted as being a right lateral
transform fault related to the opening up of the
oceans. Thus the *apparent* displacement of the ridge
is not true, but due to the shape of the split between
continents at the outset of the parting of plates.

transgression the result of an increase in sea level
producing an advance of the sea over new land
areas. Thus, deeper water sediments are deposited
over shallow water sediments.

transit movement of a small body across the disk of
a larger body (to an observer on Earth), as with a
satellite moving across its parent planet.

transition point the point at which a substance
may exist in more than one solid form, in equilibrium.

transition element an ELEMENT characterized by an
incomplete inner electron shell and a variable valency. Metallic in nature, the chemical properties of
one element resemble those of the adjacent element
in the PERIODIC TABLE.

transitive a relationship between mathematical entities such that if one object bears a relation to a
second object, which bears a relation to a third
object, then the first object bears this relationship

to the third object, e.g. if a = b and b = c then a = c.

transistor a SEMICONDUCTOR device that is used in three main ways: as a switch; a rectifier (or DIODE, which conducts current in one direction, thus turning AC into DC); and as an amplifier creating strong signals from weak ones.

translation the synthesis of PROTEINS in a RIBOSOME that has MESSENGER RNA (mRNA) attached to it. As the mRNA molecule moves through part of the ribosome, a TRANSFER RNA molecule carrying the appropriate AMINO ACID will enter a site on the ribosome and will be released after it has contributed a new amino acid to the growing chain.

transpiration the loss of water vapour from pores (stomata) in the leaves of plants. Transpiration can sometimes account for the loss of over one sixth of the water that has been taken up by the plant roots. The transpiration rate is affected by many environmental factors—temperature, light and carbon dioxide (CO_2) levels, air currents, humidity and the water supply from the plant roots. The greatest transpiration rate will occur if a plant is photosynthesizing (*see* PHOTOSYNTHESIS) in warm, dry and windy conditions.

transpose in mathematics, to move one term or element from one side of an equation to the other with a corresponding reversal in sign. A MATRIX formed from another by interchanging the rows and columns: the transpose of matrix A is usually denoted AT.

transverse wave a wave in which the vibration occurs at right angles to the direction of wave propagation, e.g. an ELECTROMAGNETIC WAVE or, more simply, a wave on a taut piece of string.

trapezium a QUADRILATERAL with two parallel sides (*see also* ISOSCELES).

tribology the study of FRICTION, lubrication and wear, as when two surfaces are in contact in relative motion. It includes the study of substances that diminish wear, overheating, etc. in such circumstances.

trigonometric function one of the functions, such as sin(x), tan(x) and cos(x), obtained from studying certain ratios of the sides of a right-angled triangle.

trigonometry the study of right-angled triangles and their TRIGONOMETRIC FUNCTIONS.

trilobites a class of primitive arthropods that became extinct in the Permian. There are almost 4000 identifiable species, and after their appearance in the Cambrian (*see* APPENDIX 5) they became common in Palaeozoic seas. The body was divided into three parts (essentially into the head, thorax and tail) and also into three lobes running along the length of the body. The trilobite eye resembles the compound eye of living arthropods. On average, trilobites were up to 10cm (4 inches) long, although exceptional specimens up to one metre did exist.

trisomy the abnormal condition in which an organism has three CHROMOSOMES rather than the normal pair for one type of chromosome. Trisomy can occur in humans and results in offspring with abnormal characteristics and shorter-than-average lifespans. One common example of trisomy is Down's syndrome, caused by the presence of three instead of two chromosomes of the number 21 type (all the other chromosomes are in normal pairs).

trophic relating to nutrition

tropism growth of a plant organ in a particular

direction due to an external stimulus, e.g. touch, light.

tropopause the top of the TROPOSPHERE. The boundary between the troposphere and the stratosphere. The altitude of the tropopause differs over short periods, from 9–12 kilometres (5.5–7.5 miles) over the poles to about 17 kilometres (10.5 miles) over the equator.

troposphere the Earth's atmosphere between the surface and the TROPOPAUSE. This layer contains most of the water vapour in the atmosphere and most of the AEROSOLS in SUSPENSION. The temperature decreases with height at approximately 6.5°C per kilometre, and it is in this layer that most weather features occur.

tsunami a giant sea wave produced by sudden large-scale movement of the sea floor, whether due to EARTHQUAKE, volcanic explosion, or enormous slides and slumps. The slides and slumps may perhaps themselves be triggered by an earthquake. The effect of the sea-floor displacement is not especially apparent in the open oceans, but the waves, which can travel at speeds of several hundred kilometres per hour, contain an immense amount of energy. Consequently, when the force of the water reaches shallows or narrow inlets, the effects are catastrophic. A susceptible area is the western coast of the Pacific (Japan, etc), where there are deep-water trenches associated with zones of earthquakes, and highly populated islands.

tufa a SEDIMENTARY ROCK, usually composed of CALCIUM CARBONATE, precipitated out of solution at springs where there may be heating and expulsion of carbon dioxide (i.e. the loss of CO_2 which causes the deposition of $CaCO_3$). Tufa tends to show a porous (almost spongy) form, and is often interbedded with sands or gravels.

tundra a zone within the periglacial areas of Earth (i.e. originally those areas that bordered Pleistocene ice sheets but now the term is used to signify environments subjected to freezing and thawing of the land) comprising treeless plains that, due to PERMAFROST conditions, have soils that are waterlogged. The dominant plants are grasses, rushes, and sedges, with some herbs and dwarf woody plants.

tungsten a hard grey metal used in alloys where its hardness and resistance to corrosion are valued. It is used in CARBIDE tools and electric lamp filaments. Tungsten carbide is almost as hard as diamond and is used extensively in abrasives.

turbidite a SEDIMENTARY ROCK type, deposited by a TURBIDITY CURRENT, which, due to the nature of the latter, is variable in thickness and extent. Turbidites are a major sediment type derived from the land and deposited on continental margins. The deposits often show a well-defined and characteristic internal division of graded sands at the base, followed by laminated sands, then laminated silty sands with ripple marks. The uppermost layers comprise fine-grained silts and muds. It is thus, overall, a fining-upwards cycle, i.e. the grain size of the sediments decreases upwards. Throughout, there are SEDIMENTARY STRUCTURES, including SOLE MARKS, and convoluted bedding. The ideal turbidite sequence of lithologies and beds is named a *Bouma sequence* after A.Bouma who first rationalized the sequence.

turbidity current a current in water that is loaded

with sediment, producing a gravity-controlled body of water and sediment (within a sea, lake, etc.) that is denser than the surrounding water. Sediments on a slope within the sea, lake or delta are disturbed, perhaps by earthquake tremors, and the current flows downslope. The flow moves rapidly on the floor at speeds up to 7 metres (23 ft) per second, and sediment is deposited at the foot of the slope or on the near-level, deep-ocean floor. The currents are usually short-lived and have the effect of depositing shallow water sediments in deep-water environments. The sediments thus formed are called TURBIDITES. In carrying suspended sediment at relatively high speeds, turbidity currents possess considerable erosional force. This has been demonstrated on numerous occasions, and none better than the flows triggered by the Grand Banks earthquake (off Newfoundland) in 1929, where the current broke through numerous submarine cables.

twilight the period of partial light after sunset or before sunrise caused by the reflected sunlight in the upper atmosphere when the Sun is below the horizon. In astronomy, it is defined as beginning when the Sun is 18° below the horizon.

twinning a CRYSTAL exhibits twinning if two or more parts are joined such that some crystallographic property (e.g. a plane, *see* CRYSTAL) is common to all parts of the twin. Often in twinned crystals, one part is reversed with respect to the other, or it is rotated about a line. If one half of a twin is a reflection of the other half, the plane dividing them is the twin-plane. Similarly, the twin-axis is the axis about which the crystal must be rotated to reach the "untwinned" state (a purely theoretical concept and not related to how the twinned form developed). There are various types of twin, and certain MINERALS exhibit twins frequently, e.g. PYRITE and FELDSPAR, the former commonly seen in hand specimens as interlocking cubes, and the latter (especially plagioclase) easily visible in THIN SECTION as repeated stripes due to *multiple* twinning.

typhoon *see* **hurricane**

U

ultracentrifuge a machine that generates high centrifugal forces as its rotor is capable of spinning at speeds of up to 50,000 revolutions per minute. The ultracentrifuge is most commonly used during the separation of the various ORGANELLES within cells. The larger and more dense organelles will form a deposit in the centrifuge tube more readily than the smaller, less dense ones. Thus, the largest organelle of any normal cell, the NUCLEUS, will be deposited at the bottom of a centrifuge tube when the ultracentrifuge spins at a force of 600g for 10 minutes, whereas the smaller MITOCHONDRION needs a higher speed of 15,000g for 5 minutes to be deposited.

ultrabasic rock an IGNEOUS ROCK with no free QUARTZ and where the silica content is less than 45 per cent. Such rocks contain a predominance of ferromagnesian minerals, eg. OLIVINE, AMPHIBOLES and PYROXENES.

ultramafic a descriptive term for igneous rocks, and partly synonymous with ULTRABASIC, referring to the very high content of ferromagnesian silicates (but little or no FELDSPAR). Often, ultramafic rocks are divided into picrites (an OLIVINE-rich basalt), peridotites (olivine and PYROXENES), and pyroxenites (when pyroxenes dominate olivine).

ultrasonic a term used to describe sound waves that

are inaudible to humans as they have a frequency above 20kHz. Although the human ear is incapable of detecting such a high FREQUENCY, some animals, such as dogs and bats, can detect ultrasonic waves (*also known as* **ultrasound**). Ultrasound is used widely in industry, medicine and research. For example, it is used to detect faults or cracks in underground pipes and to destroy kidney stones and gallstones. The most recent development in ultrasonics is their use in chemical processes to trigger reactions involved in the production of food, plastics and antibiotics. Ultrasonics make certain chemical processes safer and cheaper as they eliminate the need for high temperatures and expensive catalysts.

ultraviolet radiation a form of radiation that occurs beyond the violet end of the visible light spectrum of ELECTROMAGNETIC WAVES. Ultraviolet rays have a FREQUENCY ranging from 10^{15}Hz to 10^{18}Hz, with a wavelength ranging from 10^{-7}m to 10^{-10}m. They are part of natural sunlight and are also emitted by white-hot objects (as opposed to red-hot objects, which emit INFRARED RADIATION). As well as affecting photographic film and causing certain minerals to fluoresce, ultraviolet radiation will rapidly destroy bacteria. Although ultraviolet rays in sunlight will convert steroids in human skin to vitamin D (essential for healthy bone growth), an excess can cause irreversible damage to the skin and eyes and damage the structure of the DNA in cells by producing THYMINE-thymine DIMERS. Fortunately, a great deal of the ultraviolet radiation from the Sun does not reach the Earth as the OZONE LAYER in the upper atmosphere acts as a UV filter.

umbra a region of complete shadow, which is usually applied to eclipses. The area of half or partial shadow is termed the penumbra.

unconformity a break in the deposition of sedimentary rocks, allowing erosion of previously formed rock before eventual deposition of further sediments. It is usually represented by an obvious difference in the attitude of the rocks on either side of the unconformity, with the upper lying UNCONFORMABLY on the lower.

unified scale (*see also* RELATIVE ATOMIC MASS) the scale that lists atomic and molecular weights using the ^{12}C isotope as the basis for the scale and taking the mass of the isotope as exactly twelve. This means the atomic mass unit is 1.660×10^{-27}kg.

unit cell the smallest fragment of a CRYSTAL that will reproduce the original crystal if the unit cells are arranged in a repeating, three-dimensional pattern.

unit vector a VECTOR with a magnitude of one.

unity the number or numeral one; a quantity assuming the value of one.

universe all matter, energy and space in the cosmos.

universal gas equation *see* **gas laws**.

universal indicator a mixture of certain substances, which will change colour to reflect the changing pH of a SOLUTION. Universal indicator is available in the form of a solution or paper strip and is used as an approximate measure of the pH of a solution by using the following chart as a guide:

Colour of indicator	red	orange	yellow	green	blue	purple
pH	1 2 3	4 5 6	7	8 9	10 11 12	13 14

unsaturated a chemical term used to indicate that

a compound or solution is capable of undergoing a chemical reaction due to specific physical properties of the compound or solution. In the case of unsaturated organic compounds, the carbon atoms are unsaturated as they form double or triple bonds and are thus capable of undergoing ADDITION REACTIONS. If a SOLUTION is described as unsaturated, then it contains a lower concentration of SOLUTE dissolved in a definite amount of the SOLVENT than the concentration of solute needed to establish the EQUILIBRIUM found in a SATURATED solution.

uracil ($C_4H_4N_2O_2$) a nitrogenous base component of the NUCLEIC ACID, RNA, that has the structure of a PYRIMIDINE. During TRANSCRIPTION, uracil will always form a BASE PAIR with ADENINE of the DNA template, and during TRANSLATION, uracil will always base pair with adenine of the MESSENGER RNA (mRNA) molecule.

uranium (U) a metallic element that is radioactive and has the greatest mass of all naturally occurring elements (atomic mass of 238). Uranium has 92 protons within its nucleus and exists as three ISOTOPES, ^{238}U, ^{235}U, and ^{234}U—each of which undergoes ALPHA DECAY. Uranium will naturally disintegrate and pass through a series of other elements to form eventually a stable isotope of the element lead. When uranium is bombarded with NEUTRONS, however, it undergoes artificial disintegration to form two other heavy nuclei, releasing a very large amount of energy—this is the basic process underlying nuclear FISSION, which is used to generate energy in nuclear power stations. Uranium is not only used as a fuel in nuclear reactors but is also used in atomic bombs. Indeed, the first atomic bomb, dropped on

Hiroshima in 1945, is believed to have contained two or more small quantities of the isotope ^{235}U, which were suddenly brought together by a device and the CHAIN REACTION of nuclear fission immediately ensued.

Uranus (*see* APPENDIX 6 for physical data) the seventh planet in the SOLAR SYSTEM with an orbit between those of SATURN and NEPTUNE. The surface temperature is about −240°C, and there are five natural satellites. Uranus also has a ring system orbiting its equator.

urea an organic molecule, $CO(NH_2)_2$, that is a metabolic byproduct of the chemical breakdown of PROTEIN in mammals. In humans, 20–30 grams of urea are excreted daily in the urine, and although urea is not poisonous in itself, an excess of it in the blood implies a defective kidney, which will cause an excess of other, possibly poisonous, waste products.

uridine a molecule consisting of the nitrogenous base URACIL and the ribose sugar that is a basic unit of RNA structure when a phosphate group (H_3PO_4) is added to form a NUCLEOTIDE.

V

vaccine a modified preparation of a VIRUS or BACTERIA
that is no longer dangerous but is capable of stimulating an immune response and thus confers immunity against infection with the actual disease.
Vaccines can be administered orally or by a hypo-
dermic syringe and not effective immediately as
it takes time for the recipient's IMMUNE SYSTEM to
develop a memory for the modified virus or bacterium by producing specific ANTIBODIES.

vacuum in theory, a space in which there is no
matter. However, a perfect vacuum is unobtainable
and the term describes a gas at a very low pressure.

vacuum distillation distillation performed under
reduced pressure. Since a reduction in pressure also
reduces the boiling point of substances, it means
that certain substances, which at normal pressures
would decompose, *can* be distilled.

vacuum tube *see* diode.

vagus nerve an important part of the nervous system that arises from the brain stem and runs down
either side of the neck. The vagus nerves accompany
the major blood vessels of the neck (internal jugular
veins) and innervate the heart and other viscera in
the chest and abdominal cavities. They are partially
responsible for the control of the heart rate and
other vital functions. Sudden stimulation of the

408

vagus nerves can cause death due to immediate cardiac failure, with the victim suddenly dropping dead. Some examples of sudden death by vagal stimulation can include a blow in the solar plexus, any form of pressure on the neck, and sudden dilatation of the neck of the womb (e.g. illegal abortion).

valency the combining power or bonding potential of an ATOM or GROUP, measured by the number of hydrogen ions (H^+, valency 1) that the atom could replace or combine with. In an IONIC compound, the charge on each ion represents the valency e.g. in NaCl, both Na^+ and Cl^- have a valency of one. In COVALENT compounds, the valency is represented by the number of bonds that are formed, thus in carbon dioxide (CO_2), the carbon has a valency of 4 and oxygen 2 (*see* VALENCY ELECTRONS).

valency electrons the electrons present in the outermost shell of an atom of an element. Some elements always have the same number of valence electrons, e.g. hydrogen has one, ordinary oxygen has two, and calcium has two. The valence electrons of an atom are the ones involved in forming bonds with other atoms and are therefore shared, lost or gained when a compound or ION is formed.

valve a piece of tissue attached to the wall of a tube that restricts the flow of the blood being carried in one direction. The most important valves are the ones found in the HEART and VEINS, which prevent a backflow of blood.

Van Allen radiation belts consist of charged particles trapped in the Earth's magnetic field and forming two belts around the Earth. The lower belt occurs between about 2000 and 5000 kilometres

(1240 and 3100 miles), and its particles are derived from the Earth's atmosphere. The particles in the upper belt are from the SOLAR WIND and the belt occurs at around 20000 kilometres (about 12,400 miles). The belts were named after the American space scientist James Van Allen (1914–), who discovered them in 1958.

Van de Graaff generator a machine, named after the American physicist Robert Jemison Van de Graaff (1901–67), that continuously separates electrostatic charges and in so doing produces a very high voltage. The fundamental structure of a Van de Graaff generator consists of a hollow metal sphere supported on an insulating tube. A motor-driven belt of, say, rubber or silk, carries positive charge from an electrode at the bottom of the belt into the sphere. The sphere gradually becomes positively charged, and in some generators of this type, voltages as high as 500kV or 10,000,000 volts can be produced. When used in conjunction with high voltage X-ray tubes, large machines with elaborate electrode systems can generate electrical energy, which is used to split atoms for research purposes.

Van der Waals' forces the weak, attractive force between two neighbouring atoms. They are named after the Dutch physicist, Johannes Diderik van der Waals (1837-1923), who first discovered the phenomenon. In any atom, the electrons are continually moving and therefore have random distribution within the electron cloud of the atom. At any one moment, the electron cloud of an atom may be distorted so that a transient DIPOLE is produced. If two non-covalently bonded atoms are close enough together, the transient dipole in one atom will disturb the electron cloud of the other. This disturbance will create a transient dipole in the second atom, which will in turn attract the dipole in the first. It is the interaction between these transient dipoles that results in weak, non-specific Van der Waals' forces. Van der Waals' forces occur between all types of molecules, but they decrease in strength with increasing distance between the atoms or molecules.

vaporization *see* **evaporation**.

vapour pressure the pressure exerted by a vapour whether in a mixture of gases or by itself. Vapour pressure is usually quoted as a state of equilibrium between the vapour of a substance and its liquid form in a closed container. It will depend on temperature and the physical properties of the liquid.

variable a changing quantity that can have different values, as opposed to a constant. In the equation $y = 3x^2 + 7$, x and y are variables, whilst 3 and 7 are constants. An INDEPENDENT VARIABLE is a variable in a function that determines the value of the other variable. A DEPENDENT VARIABLE has its value determined by other variables. So in this example, x is the independent variable and y is the dependent variable.

variable star a star with a LUMINOSITY that varies with time, due in some circumstances to the star pulsating, which affects surface temperature and size, and thus the luminosity.

variation of latitude the phenomenon whereby the Earth's axis of rotation varies about an average position, due to its irregular form. The result is that latitudinal positions also undergo periodic fluctuations.

varve a lacustrine deposit, near to ice sheets, of banded clays, silts and sands. The rhythmically banded sediments were deposited annually in lakes at the edge of ice sheets. Spring meltwaters bring new loads of sediment into the lake; the coarse particles are deposited quickly while the finer particles are only deposited from SUSPENSION later in the year. Since the glacial streams would refreeze in the winter, the sediment supply would cease until the following spring. This cyclic activity produces the banded effect, and each season accounts for one pale coarse band and one dark finer band. Varve deposits are very thin, and sixty or seventy years may be accommodated in one metre of sediment.

vector any physical quantity that has both direction and magnitude. Vectors include displacement, velocity, acceleration and momentum.

In biology, the term vector represents the plasmid used to carry a DNA segment into the host's cells or the organism that acts as a mechanism for transmitting a parasitic disease, e.g. mosquitoes are vectors of malaria.

vegetative propagation a type of reproduction in which the non-sexual organs of the plant are capable of producing progeny. Vegetative propagation occurs naturally in certain plants, e.g. potato tubers and strawberry runners.

vein any thin-walled vessel that carries blood back from the body to the HEART. Veins contain few muscle fibres but have one-way VALVES that prevent backflow, thus enabling the blood to flow from body areas back to the heart.

In geology, a sheet-like feature usually occupying a fracture or fissure within a rock, which is infilled with mineral deposits. CALCITE and QUARTZ commonly form veins, but ore deposits do occur in this form, commonly mixed with other minerals.

velocity the rate of change of position of any object. Velocity (v) is a VECTOR quantity, and therefore it should be expressed in both magnitude and direction. The unit of velocity is metres per second (ms^{-1}) and can be calculated using the displacement (s) and time elapsed (t) as follows: $V = s/t$. An object is described as moving with constant velocity when it is travelling along a straight line in equal proportions of distance against time. However, it is more likely that an object's velocity changes with time, in which case the object is said to be accelerating (*see* ACCELERATION).

velocity of light the VECTOR quantity for light travelling through a given medium. The velocity of light in a vacuum and in air hardly differs and is approximately $3.0 \times 10^8 ms^{-1}$. However, the velocity of light in water is approximately $2.3 \times 10^8 ms^{-1}$, and it is this difference in velocity in air and water that explains the REFRACTION of light when passing from one medium to another (*see also* SPEED OF LIGHT).

vena cava one of the two major veins that empty into the right chamber of the HEART. The superior vena cava (SVC) carries the blood collected from the upper part of the body, e.g. neck and brain, while the inferior vena cava (IVC) carries the blood from the lower half of the body, e.g. liver, kidney and legs.

ventricle a major chamber of the HEART, which is thick-walled and muscular as it is the main pumping chamber. The outflow of the right ventricle is known as the PULMONARY ARTERY, which distributes blood to the lungs, and the outflow of the left ventri-

413

cle is called the AORTA, which distributes blood to the head and the rest of the body.

Venus (*see* APPENDIX 6 for physical data) the second planet in the SOLAR SYSTEM with its orbit between those of MERCURY and EARTH. The atmosphere is predominantly carbon dioxide, and the surface temperature is about 470°C. There is dense cloud cover with low temperature at altitude. Venus is seen in the sky as the bright EVENING and MORNING STAR. Surface features include mountain ranges, craters, and a rift valley. It has no satellites.

vertex the point at which two sides of a polygon or the planes of a solid intersect.

vesicular structure bubble-shaped cavities formed in LAVA, caused by trapped gas bubbles. The cavities are often filled subsequently with crystals deposited from solution.

virga a feature relating to certain CLOUD formations, where trails of precipitation (rain or snow) fall beneath the cloud but evaporate before reaching the ground.

virus the smallest microbe, which is completely parasitic as it is incapable of growing or reproducing outside the cells of its host. Most, but not all, viruses cause disease in plant, animal and even bacterial cells. Viruses are classified according to their nucleic acids and can contain double-stranded (DS) or single-stranded (SS) DNA or RNA. In infection, any virus must first bind to the host cells and then penetrate to release the viral DNA or RNA. The viral DNA or RNA then takes control of the cell's metabolic machinery to replicate itself, form new viruses, and then release the mature virus by either budding from the cell wall or rupturing and hence killing the cell. Some familiar examples of virus-induced diseases are herpes (double-stranded DNA), influenza (single-stranded RNA) and the retroviruses (single-stranded RNA, believed to cause AIDS and perhaps CANCER).

viscosity a property of fluids that indicates their resistance to flow. For example, oil is more viscous than water, and an object falling through oil is much slower than the same object falling through water because of the greater viscous force acting on it. A perfect fluid would be non-viscous.

vitamins organic compounds that are required in small amounts in the diet, to maintain good health. Deficiencies lead to specific diseases. Vitamins form two groups: A, D, E and K are fat-soluble, while C (ASCORBIC ACID) and B (thiamine) are water-soluble.

viviparous a term describing any animal that gives birth to young that have developed inside its body. Viviparity is not restricted to mammals but also applies to some species of insect, e.g. the species of mite, *Acarophenox*, whose young develop by devouring and thus killing the mother.

volatile a term describing any substance that can easily change from the solid or liquid state to its vapour, i.e. a substance which has a high vapour pressure and which passes into the gaseous PHASE rapidly.

In geology, elements that would ordinarily be gaseous under normal conditions are dissolved in a MAGMA because of the pressure and nature of the melt. On reaching the surface, or reduced pressure, the volatiles become gaseous, forming carbon dioxide, sulphur dioxide, hydrochloric acid, and others. Concentrations of volatiles in late melts form

PEGMATITES. Volatiles in COAL include methane and hydrogen, and water vapour. Anthracite contains about 10 per cent volatiles, which increases to approximately 50 per cent or more in peat.

volcanic ash TEPHRA that is less than 2 mm in size.

volcanic bomb a lump of LAVA that is ejected from a VOLCANO and that can take one of several shapes, depending upon the type of lava, the degree of solidification, and its flight through the air.

volcanic neck *see* **plug.**

volcanic vent the pipe that connects a volcanic crater with the magma source.

volcano a natural vent or conduit at the Earth's surface, connecting with the interior and from which a variety of solid, molten and gaseous materials are erupted. The eruptions may consist of steady flows of LAVA, explosions of ash, gas and fragments, or hot flows of ash. The type of eruption tends to be named after specific volcanoes or areas of volcanic activity, e.g. Hawaiian (outpouring of fluid basaltic lavas), Peléean (violent eruptions with NUÉES ARDENTES, and viscous lavas), Strombolian (moderate eruptions with small explosions and basaltic lava of average VISCOSITY), and Vesuvian (very explosive after long dormant periods with frothy gas-filled lava and clouds of ash and gas). Volcanoes may be active, dormant or extinct.

volt (V) the unit of POTENTIAL DIFFERENCE. One volt is equal to one joule per coulomb of charge, i.e.

$$V = JC^{-1}.$$

voltage the electrical energy that moves charge around a CIRCUIT. Voltage is the same as POTENTIAL DIFFERENCE, and is thus measured in VOLTS. It is calculated between two given points on a circuit,

and can be derived from the following equation:
$$V = d \times E,$$
where V = potential difference, d = difference between 2 points, E = electric field strength.

volume the space occupied by any object or substance. The volume of a liquid will depend on the amount of container space it occupies, but the volume of any gas will vary with pressure and temperature. Volume is measured in cm^3 or m^3. The volume of a cube, cuboid or cylinder is equal to the area of the base x height; the volume of a pyramid or cone is equal to $1/3$ of the area of the base x height. The volume of a sphere is $4/3\pi r^3$. The

volumetric analysis chemical analysis that uses standard solutions of known concentrations to calculate a particular constituent present in another solution, using TITRATION.

vulgar fraction an ordinary fraction with one number over the other, e.g. $3/5$, $5/9$, $1/16$.

vug a cavity in a rock, which is often lined or infilled with crystals.

W

warm front the edge of a mass of warm air advancing and rising over cold air. Cloud develops in the rising air, with heavy NIMBOSTRATUS forming as the front passes. There is usually heavy rain associated with the front. The temperature rises after the front, with the rain clearing and often a change in wind direction.

water a ubiquitous compound, hydrogen oxide (H_2O), which can occur as solid, liquid and gas phases. It forms a very large part of the Earth's surface and is vital to life. It occurs in all living organisms and has a remarkable combination of properties in its solvent capacity, chemical stability, thermal properties, and abundance.

water potential the tendency of water to move by diffusion, osmosis or as vapour. At a pressure of one atmosphere, pure water is given a water potential value of zero, and hence all cells that water enters by osmosis have a water potential value less than zero. The water potential of any cell can be calculated using the following:

water potential = osmotic potential + pressure potential.

water table the level below which water saturates the available spaces in the ground. A spring or river is formed when, due to geological conditions, the water table rises above ground level.

Watson, James Dewey (1928–) an American molecular biologist who, along with his colleague, the English biologist Francis H. Crick (1916–), constructed a model revealing the structure of the DNA molecule. In 1962, they shared the Nobel prize for their work, and their double-helical model of DNA, showing a simple, repeating pattern of paired nucleic acid bases, suggested a means by which DNA replicates. Watson published an account of the discovery of DNA structure in his book, *The Double Helix* (1968).

watt (W) a unit of power that is the rate of WORK done at 1 JOULE per second, i.e. $1W = 1Js^{-1}$.

wave a mechanism of energy transfer through a medium. The origin of the wave is vibrating particles, which store and release energy while their mean position remains constant as it is only the wave that travels. Waves can be classified as being either LONGITUDINAL WAVES, e.g. sound, or TRANSVERSE WAVES, e.g. light, depending on the direction of their vibrations. There is a basic wave equation that relates the wavelength (λ), frequency (f), and speed (c) of the wave as $c = f\lambda$. All forms of waves have the following properties: diffraction; interference; reflection and refraction. ELECTROMAGNETIC WAVES have all of these properties but differ from ordinary waves, such as water waves, in that they can travel through a vacuum, e.g. outer space. All travelling waves have the following equation:

$$y = y_0\sin2\pi\left(\frac{t}{T} - \frac{x}{\lambda}\right)$$

where y = displacemcent, y_0=amplitude, t = time, T = 1/f (f=frequency), x = distance, λ = wavelength.

wavelength (λ) the distance between two similar points on a wave, which have exactly the same displacement value from the rest position. An example would be the distance between two crests (maximum displacement) or two troughs (maximum displacement). Wavelength is a measure of distance and hence has units of metres (m).

way up applied particularly to areas of folding and deformation, way up is the upward direction (or correct sequence from older to younger) of a sequence of rocks. SEDIMENTARY STRUCTURES, e.g. SOLE MARKS, CROSS-BEDDING, GRADED BEDDING, are used to determine way up.

weathering a combination of physical and chemical processes at and just beneath the surface, which breaks down minerals and rocks. The action of frost widening cracks in rocks, moving water loosening particles and contributing to chemical reactions, and other actions all assist in disintegration.

weather map a graphic chart, indicating measurements of temperature, wind speed, atmospheric pressure, cloud and precipitation, that is used as a basis for forecasting.

weight the gravitational force of attraction exerted by the earth on an object. As weight is a FORCE, its unit is the NEWTON (N). The weight of any object on earth can be calculated using:

$$W = mg \quad m = \text{mass (kg)}$$
$$g = \text{gravitational constant} = 9.8\text{ms}^{-2}$$

In everyday use, the term weight really refers to the mass of a person or object.

welded tuff (*see also* IGNIMBRITE) the welded zone within an ignimbrite body, composed of glass and ash fragments fused together.

well logging *or* **wireline logging** the production of geophysical measurements along a borehole (a wireline geophysical well log), using specialized equipment that enables calculations of various parameters, e.g. rock porosities, water saturation, moveable hydrocarbons, and through this the recognition of different rock types. There are many different well logs, some recording spontaneous phenomena (e.g. natural radioactivity as in the gamma ray log) or induced responses, as with the sonic log (or formation velocity log), which emits sound and measures the time taken to reach a receiver. Other properties measured include resistivity and conductivity (both induced), borehole temperature, and spontaneous electrical currents.

Wheatstone bridge a divided electrical circuit used for measuring electrical RESISTANCE. The circuit comprises a diamond configuration of four resistances with a VOLTAGE applied between two points (between the points on the greater dimension of the diamond), and the opposite two points (the shorter dimension) bridged by a galvanometer (a device for detecting or measuring small currents). When no current flows through the galvanometer the resistances can be paired together to calculate one unknown, using a simple equation: $R1/R2 = R3/R4$.

whistler atmospheric electric noises producing whistles on a radio receiver. The effect is caused by lightning producing ELECTROMAGNETIC radiations, which follow the Earth's magnetic field force lines and are then reflected back to earth by the upper atmosphere.

white blood cell *see* **leucocyte**.

white dwarf a type of star that is very dense with a

low luminosity. They result from the explosion of stars that have used up their available hydrogen (*see* SUN). Due to their small size, their surface temperatures are high and appear white (*see* SUPERNOVA).

white light light that contains all the visible wavelengths and which can therefore be resolved into a continuous spectrum of colours.

Wilson cloud chamber *see* **cloud chamber**.

window an enclosed outcrop of rocks beneath a THRUST PLANE and exposed by erosion and where rocks above the thrust plane completely surround the window.

work an energy transfer that has the net result of moving an object. A FORCE is said to do work only when its object of application moves in the direction of the force. The work done is calculated by multiplying the force (F) by the distance(s) through which it moves, i.e. W = FS. Work is measured in the unit of energy, the JOULE. For example, if you have to pull with a force of 50 newtons to move a box 3 metres in the direction of the force (toward yourself), then work = 50N x 3.0M = 150Nm = 150J.

X

x-axis the "horizontal" axis in plane (two-dimensional) co-ordinate geometry.

X-chromosome one kind of CHROMOSOME that is involved in the sex determination of an individual. A woman has a pair of X-chromosomes, whereas a man has one X-chromosome and one Y-chromosome. There are many GENES on the X-chromosome which have nothing to do with the sex of the individual. For example, red-green colour blindness is determined by a RECESSIVE gene on the X-chromosome. If a woman carrying this gene has a son (X,Y) then he will inherit colour blindness as the Y-chromosome will have no corresponding gene to suppress the effect. If she has a daughter (X,X), and the father has normal vision, then the recessive gene is still inherited but its effect is suppressed as the X-chromosome from the father will carry the dominant gene for normal vision.

xenocryst (*compare with* PHENOCRYST) a crystal in an IGNEOUS ROCK that has been introduced from the surrounding rocks into the magma and has not crystallized from the melt. As such, xenocrysts are not usually in equilibrium with the melt.

xenolith a fragment of pre-existing SEDIMENTARY, IGNEOUS or METAMORPHIC ROCK enclosed within an igneous rock. They are often derived from the coun-

423

try (host) rocks intruded by an igneous body and can often show evidence of alteration. The size varies enormously.

xerography a copying process in which an ELECTRO-STATIC image is formed on a surface when exposed to an optical image. A powder mix of GRAPHITE and a thermoplastic resin of opposite charge to the electrostatic image is dusted on to the surface and the particles cling to the charged areas. The image is then transferred to a sheet of paper, again through use of opposite charges, and the image is fixed by heat.

X-ray crystallography *or* **X-ray diffraction** *or* **XRD crystallography** a technique used in geology to identify minerals. It involves directing a beam of X-rays at a CRYSTAL, which is diffracted (*see* DIFFRACTION) off the planes of atoms in the crystal. By repeating the procedure and then calculating the spacing between atomic planes, a representation of the crystal's structure can be determined.

X-ray fluorescence spectrometry *or* **XRF spectrometry** an analytical method for determination of a wide range of elements in bulk rock specimens. Rock samples are prepared as ground powder compressed into flat cylinders or fused into coin-like flat discs and then excited with X-ray radiation. The radiation causes the removal of an ELECTRON from an ORBITAL, and when it is replaced, the surplus energy is emitted as a characteristic, secondary X-ray. The X-ray is measured by the spectrometer, and the intensity of the radiation compared to a standard to enable concentrations to be calculated. The technique is widely used in analysing rock samples, both for major elements and certain TRACE ELEMENTS.

Concentrations of trace elements as low as 1 to 10 parts per million can be detected, although, in many instances, the quantities are higher.

Major Elements	*Trace Elements*
Na – sodium	Rb – rubidium
Mg – magnesium	Sr – strontium
Al – aluminium	Y – yttrium
Si – silicon	Nb – niobium
P – phosphorus	Zr – zirconium
K – potassium	Cr – chromium
Ca – calcium	Ni – nickel
Ti – titanium	Cu – copper
Mn – manganese	Zn – zinc
Fe – iron	Ga – gallium
	Ba – barium
	Pb – lead
	Th – thorium
	U – uranium

X-rays the part of the ELECTROMAGNETIC spectrum with a wavelength range of approximately 10^{-12} to 10^{-9}m and a frequency range of 10^{17} to 10^{21}Hz. X-rays are produced when electrons moving at high speed are absorbed by a target. The resultant waves will penetrate solids to varying degrees, dependent on the density of the solid. Hence X-rays of certain wavelengths will penetrate flesh, but not bone or other more dense materials. X-rays serve both therapeutic and diagnostic functions in medicine and are deployed in many areas of industry where inspection of hidden, inaccessible objects is necessary.

Y

y-axis the "vertical" axis in plane (two-dimensional) co-ordinate geometry.

Y-chromosome the small chromosome that carries a dominant gene for maleness. All normal males have 22 matched pairs of chromosomes and one unmatched pair, one large X-chromosome and one small Y-chromosome. The X-chromosome, which is inherited from the mother, carries many more genes than the Y-chromosome, which is inherited from the father. During sexual reproduction, the mother must contribute one X-chromosome, but the father can contribute either an X or Y-chromosome. The effect of the Y-chromosome is that a testis develops in the embryo instead of an ovary. Thus the sex of the resulting offspring is dependent on the father's contribution—female (X,X) or male (X,Y).

yeast unicellular micro-organisms that form a fungus. Yeast cells can be circular or oval in shape and reproduce by spore formation. The enzymes secreted by yeasts are used in brewing and baking industries as they can convert sugars into alcohol and carbon dioxide.

yield point HOOKE'S LAW states that for a material such as steel, in wire form, the extension is proportional to the tension, up to the elastic limit. An increase in tension beyond this limit takes the material to the yield point, where a sudden increase in elongation occurs with only a small further increase in tension.

younging, direction of *or* **facing direction** when dealing with FOLDS, the direction in which the sequence of rocks becomes younger. In an ANTICLINE the direction of younging is upwards because younger beds lie on top of the arched fold. However, in an antiform (an arching-upwards fold in rocks where younger rocks may not be uppermost) the direction of younging could be up or down, depending on the rock sequence, i.e. the STRATIGRAPHY. The use of WAY UP structures and features may assist in determining the direction of younging.

Young's modulus a method for calculating a ratio concerning the elasticity of a solid. Young's modulus (E) relates the STRESS and STRAIN in a solid (usually wire) using the following:

$$E = stress/strain = \sigma/\varepsilon$$

where σ = Force/Area and ε = Change in Length/ Length

Young's modulus has units of newtons per metre squared (Nm^{-2}) and is calculated only when the material is under elastic conditions, i.e. the applied force does not exceed the elastic limit and cause deformation.

Z

zenith the highest point—the pole—that is vertically above the observer with the latter in the centre of the horizon circle. Opposite to the NADIR in the CELESTIAL SPHERE.

zenith distance the angular distance of a heavenly body from the ZENITH.

zeolites a group of natural and synthetic hydrated alumina silicates of sodium, potassium, calcium and barium that contain loosely held water that can be removed by heating and regained by exposure to water. However, the cavities created by the loss of water can be occupied by other molecules of a similar size. Zeolites have thus found uses in ION EXCHANGE and as adsorbents. Zeolites containing small amounts of platinum or palladium are used as CATALYSTS in the CRACKING of HYDROCARBONS.

zodiac a zone within the CELESTIAL SPHERE that contains the paths of the planets, the Moon and the Sun. It is divided into the signs of the zodiac, named after the constellations.

zone a rock unit identified by the FOSSILS it contains (often called biostratigraphic zone).

zoology a branch of biology that involves the study of animals. Subjects studied include anatomy, physiology, embryology, evolution, and the geographical distribution of animals.

428

zwitterion the predominant form of an AMINO ACID when surrounded by a neutral solution (pH 7). The structure of a zwitterion and amino acid differ in that the zwitterion exists as a dipolar ion—the carboxyl (-COOH) group of the amino acid loses a hydrogen atom to form -COO⁻, and the amino group (-NH$_2$) gains a hydrogen atom to form -NH$_3$⁺.

zygote the cell immediately produced by the fusion of male and female germ cells (GAMETES) during the initial stage of FERTILIZATION. The zygote is a DIPLOID cell, formed by the fusion of the haploid male gamete and the haploid female gamete.

zymogen an inactive form of an ENZYME. Most zymogens are inactive precursors of pancreatic enzymes, which are involved in PROTEIN digestion. Synthesis of these digestive enzymes as zymogens prevents the unwanted digestion of the tissue in which the enzyme was made. The zymogen becomes activated by chemical modifications to form its functional form when it reaches its site of function, e.g. the enzyme chymotrypsin (digests protein) is synthesized in the pancreas as the zymogen, chymotrypsinogen, and becomes activated only when it reaches its destination, the small intestine.

zymurgy a branch of chemistry that involves the study of FERMENTATION processes.

Appendices

APPENDIX 1

Periodic Table

Group																	
1A	2A	3B	4B	5B	6B	7B	8	8	8	1B	2B	3A	4A	5A	6A	7A	0
H 1																	He 2
Li 3	Be 4											B 5	C 6	N 7	O 8	F 9	Ne 10
Na 11	Mg 12	<									>	Al 13	Si 14	P 15	S 16	Cl 17	Ar 18
K 19	Ca 20	Sc 21	Ti 22	V 23	Cr 24	Mn 25	Fe 26	Co 27	Ni 28	Cu 29	Zn 30	Ga 31	Ge 32	As 33	Se 34	Br 35	Kr 36
Rb 37	Sr 38	Y 39	Zr 40	Nb 41	Mo 42	Tc 43	Ru 44	Rh 45	Pd 46	Ag 47	Cd 48	In 49	Sn 50	Sb 51	Te 52	I 53	Xe 54
Cs 55	Ba 56	La¹ 57	Hf 72	Ta 73	W 74	Re 75	Os 76	Ir 77	Pt 78	Au 79	Hg 80	Tl 81	Pb 82	Bi 83	Po 84	At 85	Rn 86
87	88	89															

← — — TRANSITION ELEMENTS — — →

¹ Lanthanides	La 57	Ce 58	Pr 59	Nd 60	Pm 61	Sm 62	Eu 63	Gd 64	Tb 65	Dy 66	Ho 67	Er 68	Tm 69	Yb 70	Lu 71
² Actinides	Ac 89	Th 90	Pa 91	U 92	Np 93	Pu 94	Am 95	Cm 96	Bk 97	Cf 98	Es 99	Fm 100	Md 101	No 102	Lr 103

433

APPENDIX 2

Element Table

Element	Symbol	Atomic Number	Relative Atomic Mass*
Actinium	Ac	89	{227}
Aluminium	Al	13	26.9815
Americium	Am	95	{243}
Antimony	Sb	51	121.75
Argon	Ar	18	39.948
Arsenic	As	33	74.9216
Astatine	At	85	{210}
Barium	Ba	56	137.34
Berkelium	Bk	97	{247}
Beryllium	Be	4	9.0122
Bismuth	Bi	83	208.98
Boron	B	5	10.81
Bromine	Br	35	79.904
Cadmium	Cd	48	112.40
Caesium	Cs	55	132.905
Calcium	Ca	20	40.08
Californium	Cf	98	{251}
Carbon	C	6	12.011
Cerium	Ce	58	140.12
Chlorine	Cl	17	35.453
Chromium	Cr	24	51.996
Cobalt	Co	27	58.9332
Copper	Cu	29	63.546
Curium	Cm	96	{247}
Dysprosium	Dy	66	162.50

Element	Symbol	Atomic Number	Relative Atomic Mass*
Einsteinium	Es	99	{254}
Erbium	Er	68	167.26
Europium	Eu	63	151.96
Fermium	Fm	100	{257}
Fluorine	F	9	18.9984
Francium	Fr	87	{223}
Gadolinium	Gd	64	157.25
Gallium	Ga	31	69.72
Germanium	Ge	32	72.59
Gold	Au	79	196.967
Hafnium	Hf	72	178.49
Helium	He	2	4.0026
Holmium	Ho	67	164.930
Hydrogen	H	1	1.00797
Indium	In	49	1114.82
Iodine	I	53	126.9044
Iridium	Ir	77	192.2
Iron	Fe	26	55.847
Krypton	Kr	36	83.80
Lanthanum	La	57	138.91
Lawrencium	Lr	103	{257}
Lead	Pb	82	207.19
Lithium	Li	3	6.939
Lutetium	Lu	71	174.97
Magnesium	Mg	12	24.305
Manganese	Mn	25	54.938
Mendelevium	Md	101	{258}

Element	Symbol	Atomic Number	Relative Atomic Mass*
Mercury	Hg	80	200.59
Molybdenum	Mo	42	95.94
Neodymium	Nd	60	144.24
Neon	Ne	10	20.179
Neptunium	Np	93	{237}
Nickel	Ni	28	58.71
Niobium	Nb	41	92.906
Nitrogen	N	7	14.0067
Nobelium	No	102	{255}
Osmium	Os	76	190.2
Oxygen	O	8	15.9994
Palladium	Pd	46	106.4
Phosphorus	P	15	30.9738
Platinum	Pt	78	195.09
Plutonium	Pu	94	{244}
Polonium	Po	84	{209}
Potassium	K	19	39.102
Praseodymium	Pr	59	140.907
Promethium	Pm	61	{145}
Protactinium	Pa	91	{231}
Radium	Ra	88	{226}
Radon	Rn	86	{222}
Rhenium	Re	75	186.20
Rhodium	Rh	45	102.905
Rubidium	Rb	37	85.47
Ruthenium	Ru	44	101.07
Samarium	Sm	62	150.35

Element	Symbol	Atomic Number	*Relative Atomic Mass**
Scandium	Sc	21	44.956
Selenium	Se	34	78.96
Silicon	Si	14	28.086
Silver	Ag	47	107.868
Sodium	Na	11	22.9898
Strontium	Sr	38	87.62
Sulphur	S	16	32.064
Tantalum	Ta	73	180.948
Technetium	Tc	43	{97}
Tellurium	Te	52	127.60
Terbium	Tb	65	158.924
Thallium	Tl	81	204.37
Thorium	Th	90	232.038
Thulium	Tm	69	168.934
Tin	Sn	50	118.69
Titanium	Ti	22	47.90
Tungsten	W	74	183.85
Uranium	U	92	238.03
Vanadium	V	23	50.942
Xenon	Xe	54	131.30
Ytterbium	Yb	70	173.04
Yttrium	Y	39	88.905
Zinc	Zn	30	65.37
Zirconium	Zr	40	91.22

*Values of the *Relative Atomic Mass* in brackets refer to the most stable, known, isotope.

Elements listed by symbol

Symbol	Element	Symbol	Element
Ac	Actinium	Mn	Manganese
Ag	Silver	Mo	Molybdenum
Al	Aluminium	N	Nitrogen
Am	Americium	Na	Sodium
Ar	Argon	Nb	Niobium
As	Arsenic	Nd	Neodymium
At	Astatine	Ne	Neon
Au	Gold	Ni	Nickel
B	Boron	No	Nobelium
Ba	Barium	Np	Neptunium
Be	Beryllium	O	Oxygen
Bi	Bismuth	Os	Osmium
Bk	Berkelium	P	Phosphorus
Br	Bromine	Pa	Protactinium
C	Carbon	Pb	Lead
Ca	Calcium	Pd	Palladium
Cd	Cadmium	Pm	Promethium
Ce	Cerium	Po	Polonium
Cf	Californium	Pr	Praseodymium
Cl	Chlorine	Pt	Platinum
Cm	Curium	Pu	Plutonium
Co	Cobalt	Ra	Radium
Cr	Chromium	Rb	Rubidium
Cs	Caesium	Re	Rhenium
Cu	Copper	Rh	Rhodium
Dy	Dysprosium	Rn	Radon
Er	Erbium	Ru	Ruthenium
Es	Einsteinium	S	Sulphur
Eu	Europium	Sb	Antimony
F	Fluorine	Sc	Scandium
Fe	Iron	Se	Selenium
Fm	Fermium	Si	Silicon
Fr	Francium	Sm	Samarium
Ga	Gallium	Sn	Tin
Gd	Gadolinium	Sr	Strontium
Ge	Germanium	Ta	Tantalum
H	Hydrogen	Tb	Terbium
He	Helium	Tc	Technetium
Hf	Hafnium	Te	Tellurium
Hg	Mercury	Th	Thorium
Ho	Holmium	Ti	Titanium
I	Iodine	Tl	Thallium
In	Indium	Tm	Thulium
Ir	Iridium	U	Uranium
K	Potassium	V	Vanadium
Kr	Krypton	W	Tungsten
La	Lanthanum	Xe	Xenon
Li	Lithium	Y	Yttrium
Lr	Lawrencium	Yb	Ytterbium
Lu	Lutetium	Zn	Zinc
Md	Mendelevium	Zr	Zirconium
Mg	Magnesium		

APPENDIX 3

THE GREEK ALPHABET

Name	Capital	Lower Case	English Sound
alpha	A	α	a
beta	B	β	b
gamma	Γ	γ	g
delta	Δ	δ	d
epsilon	E	ε	e
zeta	Z	ζ	z
eta	H	η	e
theta	Θ	θ	th
iota	I	ι	i
kappa	K	κ	k
lambda	Λ	λ	l
mu	M	μ	m
nu	N	ν	n
xi	Ξ	ξ	x
omicron	O	ο	o
pi	Π	π	p
rho	P	ρ	r
sigma	Σ	σ	s
tau	T	τ	t
upsilon	Y	υ	u
phi	Φ	φ	ph
chi	X	χ	kh
psi	Ψ	ψ	ps
omega	Ω	ω	o

The International System of Units (SI units)

Quantity	Symbol	Unit	Symbols
acceleration	a	metres per second squared	ms^{-2} or m/s^2
area	A	square metre	m2
capacitance	C	farad	F (1F = 1AsV^{-1})
charge	Q	coulomb	C (1C = 1As)
current	I	ampere	A
density	ρ	kilograms per cubic metre	kgm^{-3} or kg/m^3
force	F	newton	N (1N = 1 kg ms^{-2})
frequency	f	hertz	Hz (1Hz = 1s^{-1})
length	l	metre	m
mass	m	kilogram	kg
potential difference	V	volt	V (1V = 1JC^{-1} or WA^{-1})
power	P	watt	W (1W = 1Js^{-1})
resistance	R	ohm	Ω (1Ω = 1VA^{-1})
specific heat capacity	c	joules per kilogram kelvin	Jkg^{-1} K^{-1}
temperature	T	kelvin	L
time	t	second	s
volume	V	cubic metre	m^3
velocity	v	metres per second	ms^{-1} or m/s
wavelength	λ	metre	m
work, energy	W, E	joule	J (1J = 1Nm)

APPENDIX 4 (cont.)
Useful prefixes adopted with SI units

Prefix	Symbol	Factor
tera	T	10^{12}
giga	G	10^{9}
mega	M	10^{6}
kilo	k	10^{3}
hecto	h	10^{2}
deda	da	10^{1}
deci	d	10^{-1}
centi	c	10^{-2}
milli	m	10^{-3}
micro	μ	10^{-6}
nano	n	10^{-9}
pico	p	10^{-12}
femto	f	10^{-15}
atto	a	10^{-18}

Geological Time Scale

Eon	Era	Sub-era	Period	Epoch	Millions of years since the start
PHANEROZOIC	Cenozoic	Quaternary	Pleistogene	Holocene	0.01
				Pleistocene	2.0
		Tertiary	Neogene	Pliocene	5.1
				Miocene	24.6
			Palaeogene	Oligocene	38
				Eocene	55
				Palaeocene	65
	Mesozoic		Cretaceous		144
			Jurassic		213
			Triassic		248

Geological Time Scale

Eon	Era	Sub-era	Period	Epoch	Millions of years since the start
PHANEROZOIC	Palaeozoic	Upper Palaeozoic	Permian		286
			Carboniferous		360
			Devonian		408
		Lower Palaeozoic	Silurian		438
			Ordovician		505
			Cambrian		590
PROTEROZOIC P R E					2500
C A M B ARCHAEAN					4000
R I A N PRISCOAN					4600

The Solar System

Planet	Diameter at the Equator km	Mass relative to the Earth[1]	Average distance from Sun km[6]	The planet's "year"
Mercury	44840	0.054	57.91	87.969 days
Venus	12300	0.8150	108.21	224.701 days
Earth	12756	1.000	149.60	365.256 days
Mars	6790	0.107	227.94	686.980 days
Jupiter	142700	317.89	778.34	11.86 years
Saturn	120800	95.14	1427.01	29.46 years
Uranus	50800	14.52	2869.6	84.0 years
Neptune	48600	17.46	4496.7	164.8 years
Pluto	3500	0.1 (approx)	5907	248.4 years
Sun	1392000	332 958		
Moon	3476	0.0123		

[1] The mass of the Earth is 5.976×10^{24} kg